Cross-Platform UIs with Flutter

Unlock the ability to create native multiplatform UIs using a single code base with Flutter 3

Ryan Edge

Alberto Miola

BIRMINGHAM—MUMBAI

Cross-Platform UIs with Flutter

Group Product Manager: Rohit Rajkumar

Publishing Product Manager: Ashitosh Gupta

Senior Editor: Hayden Edwards

Content Development Editor: Abhishek Jadhav

Technical Editor: Saurabh Kadave

Copy Editor: Safis Editing

Project Coordinator: Rashika Ba

Proofreader: Safis Editing

Indexer: Hemangini Bari

Production Designer: Joshua Misquitta

Marketing Coordinator: Elizabeth Varghese and Teny Thomas

First published: August 2022

Production reference: 1240822

Published by Packt Publishing Ltd.

Livery Place

35 Livery Street

Birmingham

B3 2PB, UK.

ISBN 978-1-80181-049-4

www.packt.com

To my parents, Judy and Stephen Edge, for their constant support and for

exemplifying love, faith, and determination. To my wife Erica, and children,

Amaya, Tristan, and Bryce, for putting up with all of my shenanigans.

– Ryan Edge

To my beloved Giorgia, my family, and all my friends.

– Alberto Miola

Contributors

About the authors

Ryan Edge is an experienced software engineer, with over 10 years of experience as a web and mobile developer. He graduated in computer science from Southern Polytechnic State University. He is currently working for a stealth start-up and part-time as a freelancer, with over 3 years of professional experience in Flutter. He is a Google Developer Expert in Flutter, an active member of the open source community, and a co-organizer of his local Flutter meetup group.

Alberto Miola is an Italian software engineer who graduated in computer science from the University of Padua. He's currently working with Dart and Flutter, with which he has more than 3 years of professional experience, and also is a Dart and Flutter GDE. He attends online conferences, writes technical articles about Flutter, and is also the author of the *Flutter Complete Reference* book series.

About the reviewers

Amit Bhave is an experienced full stack developer, having worked on various languages and frameworks such as Flutter, Java, Kotlin, Spring Boot, Micronaut, and so on. He currently works as a senior engineer at Getir, a Turkish delivery services company. Previously, he has also worked at ThoughtWorks.

Ivan Rendulić is an experienced software developer and architect with more than 20 years in the software industry. He started programming in the fifth grade and is still enjoying working with and exploring new and progressive technology. He is currently working for his own consulting and development company, Butterfly Design (Croatia).

He fell in love with Flutter at first sight. He considers himself a passionate Flutter enthusiast. He has led and developed greenfield Flutter projects for clients in Croatia. Here are a couple that he is most proud of:

- Mobility One, Mobility One project, Croatia
- Dream Agency, Kali Sara App project, Croatia

He wants to pass on his full regards to his colleagues and associates:

- Alperen Yalcin (Turkey, alper50)
- Carlos Eduardo De Oliveira (Brazil, github-carlos)
- David Machara (Botswana, SidneyMachara)

Also, he wants to pass on his regards to the Flutter Croatia Community:

- Marko Filipović (Croatia, markfili)
- Sandro Lovnički (Croatia, slovnicki)

Table of Contents

6

Building a Simple Contact Application with Forms and Gestures

7

Building an Animated Excuses Application

8

Build an Adaptive, Responsive Note-Taking Application with Flutter and Dart Frog

9

Writing Tests and Setting Up GitHub Actions

Preface

Flutter is a UI toolkit for building beautiful, natively compiled applications for mobile, web, desktop, and embedded devices from a single code base. With Flutter, you can write your code once and run it anywhere using a single code base to target multiple platforms. This book is a comprehensive, project-based guide for new and emerging Flutter developers that will help empower you to build bulletproof applications.

Once you've picked up the book, you'll quickly realize what sets Flutter apart from its competition and establish some of the fundamentals of the toolkit. As you work on various project applications, you'll understand just how easy Flutter is to use for building stunning user interfaces. The book covers navigation strategies, state management, advanced animation handling, and the two main UI design styles: Material and Cupertino. It'll help you extend your knowledge with good code practices, UI testing strategies, and CI setup to constantly keep your repository's quality at the highest level possible.

By the end of this book, you'll feel confident in your ability to transfer the lessons from the example projects and build your own Flutter applications for any platform you wish.

Who this book is for

This book is for software developers with a good grasp of Flutter who want to learn best practices and techniques for building clean, intuitive UIs, using a single code base, for mobile and the web. Prior experience with Flutter, Dart, and **object-oriented programming** (OOP) will help you understand the concepts covered in the book.

What this book covers

Chapter 1, Building a Counter App with History Tracking to Establish Fundamentals, is about Flutter's core concepts. This chapter explores widgets, elements, render objects, and the rebuild system.

Chapter 2, Building a Race Standings App, is about layout and responsiveness. The project in this chapter covers internationalization, responsiveness, and layout strategies for high-quality applications.

Chapter 3, Building a Todo Application Using Inherited Widgets and Provider, is about sharing data in an application. You will learn about property forwarding and how to create inherited widgets.

Chapter 4, Building a Native Settings Application Using Material and Cupertino Widgets, is about building an application that looks native to iOS or Android. You will learn how to use Flutter's Material and Cupertino widgets.

Chapter 5, Exploring Navigation and Routing with a Hacker News Clone, is about handling transitions between a multi-screen application. You will learn how to use Navigator, an imperative routing mechanism, and GoRouter, a declarative routing mechanism.

Chapter 6, Building a Simple Contact Application with Forms and Gestures, is about capturing user input and turning that input into application data. You will learn how to use Flutter's Form and Gesture components.

Chapter 7, Building an Animated Excuses Application, teaches you how to build smooth animations in an application designed to give us random excuses to miss work.

Chapter 8, Build an Adaptive, Responsive Note-Taking Application with Flutter and Dart Frog, is about adapting an application to behave differently on different platforms and display information differently in different screen sizes. You will learn how to use Flutter's adaptive and responsive APIs.

Chapter 9, Writing Tests and Setting up GitHub Actions, is about testing Flutter apps and creating a CI pipeline to run tests. You will learn how to use GitHub actions along with our Flutter project.

To get the most out of this book

You will need to install Flutter on your machine, which already includes the Dart SDK, so you will get both with a single installation. We created and tested our project with Flutter 3, so make sure to use any version greater than this.

Software/hardware covered in the book	Operating system requirements
Flutter 3.0.4	Windows, macOS, or Linux
Dart 2.17.5	

If you are using the digital version of this book, we advise you to type the code yourself or access the code from the book's GitHub repository (a link is available in the next section). Doing so will help you avoid any potential errors related to the copying and pasting of code.

Download the example code files

You can download the example code files for this book from GitHub at `https://github.com/PacktPublishing/Cross-Platform-UIs-with-Flutter`. If there's an update to the code, it will be updated in the GitHub repository.

We also have other code bundles from our rich catalog of books and videos available at `https://github.com/PacktPublishing/`. Check them out!

Download the color images

We also provide a PDF file that has color images of the screenshots and diagrams used in this book. You can download it here: `https://packt.link/e2h8M`.

Conventions used

There are a number of text conventions used throughout this book.

`Code in text`: Indicates code words in text, database table names, folder names, filenames, file extensions, pathnames, dummy URLs, user input, and Twitter handles. Here is an example: "Next, open the `pubspec.yaml` file and make sure to have these two dev dependencies installed:"

A block of code is set as follows:

```
dev_dependencies:
  # https://pub.dev/packages/dart_code_metrics
  dart_code_metrics: ^4.9.1
  # https://pub.dev/packages/flutter_lints
  flutter_lints: ^1.0.4
```

When we wish to draw your attention to a particular part of a code block, the relevant lines or items are set in bold:

```
class TodosApp extends StatelessWidget {
  const TodosApp({
    Key? key,
    required this.todoController,
  }) : super(key: key);
  final TodosController todoController;
```

Bold: Indicates a new term, an important word, or words that you see onscreen. For instance, words in menus or dialog boxes appear in **bold**. Here is an example: "Create a new Flutter project in your favorite IDE and make sure to enable web support by clicking on the **Add Flutter web support** checkbox."

> **Tips or Important Notes**
> Appear like this.

Get in touch

Feedback from our readers is always welcome.

General feedback: If you have questions about any aspect o-f this book, email us at `customercare@packtpub.com` and mention the book title in the subject of your message.

Errata: Although we have taken every care to ensure the accuracy of our content, mistakes do happen. If you have found a mistake in this book, we would be grateful if you would report this to us. Please visit www.packtpub.com/support/errata and fill in the form.

Piracy: If you come across any illegal copies of our works in any form on the internet, we would be grateful if you would provide us with the location address or website name. Please contact us at copyright@packt.com with a link to the material.

If you are interested in becoming an author: If there is a topic that you have expertise in and you are interested in either writing or contributing to a book, please visit authors.packtpub.com.

Share your thoughts

Once you've read *Cross Platform UIs with Flutter*, we'd love to hear your thoughts! Scan the QR code below to go straight to the Amazon review page for this book and share your feedback.

https://www.amazon.in/review/create-review/?asin=1801810494&

Your review is important to us and the tech community and will help us make sure we're delivering excellent quality content.

1

Building a Counter App with History Tracking to Establish Fundamentals

When you decided to buy this book, you had probably already played with Flutter a bit or even already worked with it. We have organized this book in a way that chapters gradually increase in difficulty, and they can be read in any order. Even if you aren't a Flutter master, you will still be able to get through all the chapters, thanks to in-depth analysis, images, and code snippets. When you arrive at the end of the book, you'll be able to build Flutter apps up to version 2.5 in a professional and performant way.

When you create a new Flutter project, regardless of whether you're using Android Studio or **Visual Studio Code (VS Code)**, a new counter app is created for you. This is the default template used by the framework to set up a simple app you can run on mobile, desktop, and the web. In the first chapter, we'll be building an enhanced version of the counter app that also keeps a history of the increased values. Even if it may seem a trivial app to build, you'll see that there are many considerations to make.

In this chapter, we will mainly touch on three areas, as follows:

- Understanding the foundations – widgets, elements, and `RenderObjects`
- Setting up the project
- Creating an enhanced counter app

You may be surprised to discover how a simple app actually requires various bits of knowledge to be efficiently created. Mastering constant widgets, elements, state, and much more is the key to success. We're first reviewing some important theories about Flutter's foundations, and then we will start coding. Without further ado, let's dive in!

Technical requirements

Flutter runs on Windows, macOS, and Linux, so you don't really need to worry about the operating system. Make sure to have your Flutter version updated to the latest version in the `stable` channel and create a new project using—preferably—Android Studio or VS Code. If you're brave enough, there is also the possibility of coding on a simple text editor and then manually building and running projects, but this is quite inconvenient. To make sure that your Flutter environment is correctly set up, you can run the `flutter doctor` command in the console and check the output.

If you're looking for a complete Flutter installation guide, make sure to check the official documentation at `https://flutter.dev/docs/get-started/install`.

For now, we will ignore the `test` folder, but I promise we will come back to it in *Chapter 9*, *Writing Tests and Setting Up GitHub Actions*, where we will test our app and publish it to GitHub.

To keep the code clear and concise, we have removed colors and styling portions from the various snippets across the chapter. The complete code can be found at `https://github.com/PacktPublishing/Cross-Platform-UIs-with-Flutter/tree/main/chapter_1`. You will find two folders because we're first building a good but improvable version of our project, and then a better version later.

Understanding the foundations – widgets, elements, and render objects

Before moving to the project creation, we want to review some fundamental concepts of the Flutter framework: `Widget`, `Element`, and `RenderObject` types.

While a widget tree is created and managed by the developer, the Flutter framework builds and manages two other trees in parallel, called an element tree and a render object tree. At a very practical level, these three trees are used to build **user interfaces** (**UIs**) and decide when it's time to refresh them. At the highest level, there are widgets, and they come in two flavors, as outlined here:

- `StatelessWidget`: This kind of widget doesn't require a mutable state and is best used to describe those parts of the UI that never change. Stateless widgets are immutable.
- `StatefulWidget`: This kind of widget has a mutable state and is generally used when the developer needs to control the widget's life cycle or dynamic contents. Stateful widgets are immutable too.

Note that both kinds of widgets in Flutter are immutable because all of their parameters—if any—are `final`. A `StatefulWidget` itself is immutable but its *state* is mutable, and it's represented by a separated `State<T>` object. Whenever the UI needs to be updated, Flutter follows this logic:

- If a widget is initialized with a `const` constructor, then nothing special happens and the framework just skips to the next child. This is one of the main reasons why you should try to use constant constructors as much as possible.

- If a widget isn't initialized with a `const` constructor, then it's removed and regenerated with a new one. This is totally fine; Flutter has been built for this exact purpose and is really fast at replacing widgets.

When Flutter rebuilds a widget, it uses the `Element` object associated with that widget to compare the old and new instances. In case of changes, the UI is updated; otherwise, nothing happens and the framework proceeds to visit the next child.

> **Note**
>
> Don't think that `StatelessWidget` is less efficient than `StatefulWidget`. They are both subclasses of `Widget` and they're treated in the same way! The only difference is that a stateful widget has a `State` class too, which *survives* to rebuilds because it's stored in the `Element` object behind that widget.

In general, `StatefulWidget` is used when the widget needs a one-time initialization or it has to change whenever the configuration updates. The `State` class has a few useful methods for this purpose, as outlined here:

- `void initState()`: This method is called only once when the widget is created—it's like the *constructor* of a widget.

- `void didChangeDependencies()`: This is called right after `initState()` and it's used to perform initialization based on an `InheritedWidget` above it.

- `void didUpdateWidget(covariant T oldWidget)`: This is called whenever the widget configuration changes. For example, this method is invoked whenever a widget's parameter value is changed.

- `void dispose()`: This is called when the widget is removed from the tree—it's like the *destructor* of a widget.

If you don't need to use any of the preceding widget life cycle methods and don't use `setState()`, then `StatelessWidget` is the best choice.

Having three trees allows Flutter to only repaint the UI when it's really needed. Element objects are the glue between the configuration we want to create (widgets) and the actual implementation in the UI (render objects). They detect widget changes and decide whether a render object has to simply be updated *or* removed and recreated. Let's see a concrete example here:

```
Widget build(BuildContext context) {
  return  Container(
    width: 250,
    height: 250,
    color: Colors.white,
    child: const Text('I love Flutter!'),
  );
}
```

When it's time to render the UI, Flutter traverses the widget tree and calls the build method of the various stateful or stateless widgets. At the same time, it builds the element and the render tree by calling createElement() and createRenderObject(), respectively, on the widget being built. The actual situation that the framework is dealing with in the end is shown in the following screenshot:

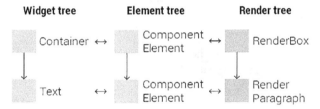

Figure 1.1 – The widget, element, and render trees

Here's what's actually happening under the hood:

- Every widget has an associated Element object that holds the state (in the case of stateful widgets) and is used by the framework to detect changes on the widget that might require a repaint of the UI.

- Elements are backed by a RenderObject that contains all the logic to paint graphics on the screen and is relatively expensive to create. It takes care of hit testing, painting, layout constraints, and much more.

- When a widget changes (using setState(), for example), the underlying element is marked as *dirty* and the framework triggers the build method to be called. Elements compare the differences between the old and the newly modified widget to update or recreate the render box.

- Render boxes are modified and kept in the framework's cache as much as possible. They are recreated only when the widget type is different. They're generally recreated only when the widget type changes or there are too many variations to apply. Updating or modifying a render object also updates the UI.

- The `RenderObject` is the one that actually has instructions to paint the UI. The Flutter framework uses elements to decide which changes, if any, have to be applied to a render object.

When you use the `const` constructor on a widget, you can save `Element` comparisons, `RenderObject` modifications, or (even better) recreations, plus other compiler and framework optimizations. Flutter already knows about the widget's location on the tree; the `BuildContext` variable is nothing more than the framework giving you access to the element behind your widget.

On the other hand, functions are *always* evaluated. When we put a function directly inside the `build` method, the returned widget (along with the `Element` and `RenderObject` types) will always be re-evaluated. Functions cannot be more efficient than an optimized, constant class hierarchy, however.

Here's a small comparison of the two widget-creation approaches, with the advantages they bring.

The only advantage functions give you is that we have less code than if creating a new class extending `StatelessWidget` or `StatefulWidget`.

Widgets instead have a lot more advantages, such as the following:

- They can have a constant constructor and allow for granular rebuilds.

- They can have keys.

- Flutter can skip the rebuild of a widget if it didn't change by looking at its `Element` object.

- They're well integrated into the Object Inspector, which may be essential in certain debugging scenarios.

- Hot reload always works as expected.

- There's no risk of reusing some previous state because they correctly dispose of all the resources.

- Since they have an `Element` object behind them, you can use the `BuildContext` object and its **application programming interface (API)**.

- They follow the **single-responsibility principle (SRP)** and lead to more maintainable, extensible, and testable code.

As you can see, preferring widgets over functions is a great idea because you can get so many advantages for free. The only optimization you need to do is using `const` in front of the constructor's name, and Flutter will take care of the rest.

Let's now jump into the practical side of things, where you see how this theory can be useful when building an efficient Flutter app!

Setting up the project

To really get started, create a new Flutter project with your favorite **integrated development environment** (**IDE**) to get the skeleton of a minimal, working Flutter app. Next, open the `pubspec.yaml` file and make sure to have these two dev dependencies installed:

```
dev_dependencies:
  # https://pub.dev/packages/dart_code_metrics
  dart_code_metrics: ^4.9.1

  # https://pub.dev/packages/flutter_lints
  flutter_lints: ^1.0.4
```

We're now going to dedicate some time to work on the `analysis_options.yaml` file. Very simply, this file contains a series of static analysis rules to help you with writing readable, high-quality code. By default, a new Flutter project already creates this file for you with a minimal setup (we have removed comments for simplicity), as illustrated in the following code snippet:

```
include: package:flutter_lints/flutter.yaml

linter:
  rules:
    avoid_print: false
    prefer_single_quotes: true
```

The `flutter_lints` package included in the file adds a series of static analysis rules recommended by the Flutter team. While this setup is fine, we suggest you improve it with stricter rules and static analysis parameters in order to do the following:

- Make sure to adhere to the Dart and Flutter standard *coding* practices.
- Improve your code quality even more.
- Increase the possibility of finding potential bugs even before compiling your code.

We get the first point *for free* because a new Flutter project already generates the minimal static analysis setup for us. For the other two points, we can use a well-known package called `dart_code_metrics`, which adds even more code metrics and static analyzer rules to the Flutter analysis machinery. After adding it to our `pubspec` file's `dev-dependencies` list, we can install it in this way:

```
include: package:flutter_lints/flutter.yaml

analyzer:
```

```yaml
plugins:
  - dart_code_metrics

dart_code_metrics:
  metrics:
    cyclomatic-complexity: 20
    number-of-parameters: 4
    maximum-nesting: 5
  metrics-exclude:
    - test/**
    - build/**
  rules:
    - avoid-unused-parameters
    - avoid-wrapping-in-padding
    - binary-expression-operand-order
    - double-literal-format
    - newline-before-return
    - no-boolean-literal-compare
    - no-equal-then-else
    - no-object-declaration
    - prefer-trailing-comma

linter:
    avoid_print: false
    prefer_single_quotes: true
    # … plus a lot more rules!
```

This is the setup we recommend for all your Flutter and Dart projects. If you wish, you can add even more rules to the `dart_code_metrics` analyzer! Just navigate to its `pub.dev` page and make sure to visit the documentation for a complete overview of the available rules.

> **Note**
>
> Each project we have hosted on GitHub contains the complete version of the `analysis_options.yaml` file we'd recommend you to use. You can easily copy and paste it into your projects from there!

Adding all of the possible linter rules to your analysis options file may work, but sometimes it doesn't make much sense. In the previous code snippet, we've tried to build a file with the most important rules, according to our experience, to help you as much as possible with Dart and Flutter's best coding practices.

At this point, we're all set, and we can start creating a counter app!

Creating an enhanced counter app

Since this chapter focuses on performance considerations and Flutter's fundamentals, we want to keep the UI minimal. The app we're going to build is just some text, two buttons, and a horizontal list view. Very intuitively, the plus and minus buttons respectively increase and decrease the text at the center by 1. New tiles are only added to the scrollable list underneath when the value is increased; tapping on minus won't add a new item in the ListView widget. You can see a representation of the UI in the following screenshot:

Figure 1.2 – The UI of our enhanced counter app

We can immediately see that the UI is made up of the following three main parts:

- A title at the top with two red and green signs
- The counter itself in the middle
- A scrollable list at the bottom

These three main UI pieces can be mapped to three widgets we need to create for the app. We have decided to structure the contents of the lib folder like so:

```
lib/
  - counter_app/
    - widgets/
        - app_title.dart
        - history.dart
    - counter_app_body.dart
  - main.dart
```

The counter_app_body.dart file contains the entire UI, which is actually made up of the three previously mentioned pieces.

As always, the main.dart file is located at the root of the lib folder, and it looks like this:

```
void main() {
  // Running the app
  runApp(
    const EnhancedCounterApp(),
  );
}


/// The root widget of the app.
class EnhancedCounterApp extends StatelessWidget {
  /// Creates an [EnhancedCounterApp] instance.
  const EnhancedCounterApp({
    Key? key,
  }) : super(key: key);

  @override
  Widget build(BuildContext context) {
    return const MaterialApp(
      home: CounterAppBody(),

      // Hiding the debug banner
      debugShowCheckedModeBanner: false,
    );
  }
}
```

Notice how we've put in the effort to document and comment all of the code. While this is not compulsory, it's highly recommended (and also enforced by the linter) that you *at least* document the public code.

Now that the app setup is finally completed, we can start creating a `CounterAppBody` widget, which is the central point of this chapter. We're going to fill this skeleton in the next section by replacing the comments in the `children` list with an actual widget, but before we do that, here's the code we need:

```
/// The contents of the counter app.
class CounterAppBody extends StatelessWidget {
  /// Creates a [CounterAppBody] widget.
  const CounterAppBody({
    Key? key,
  }) : super(key: key);

  @override
  Widget build(BuildContext context) {
    return Scaffold(
      body: Center(
        child: Column(
          mainAxisSize: MainAxisSize.min,
          children: [
            // Title widget

            // The counter widget

            // The history widget
          ],
        ),
      ),
    );
  }
}
```

In the next section, we're going to replace the inline comments with actual, useful widgets.

The title widget

The app title is the first widget we need to build. Since it's never going to change, we can safely use a stateless widget and use a constant constructor to prevent unneeded rebuilds, as follows:

```
/// This widget simply contains the title of the app.
class AppTitle extends StatelessWidget {
```

```dart
/// Creates an [AppTitle] widget.
const AppTitle({
  Key? key,
}) : super(key: key);

@override
Widget build(BuildContext context) {
  return Row(
    mainAxisSize: MainAxisSize.min,
    children: const [
      Icon(
        Icons.remove,
        color: Colors.redAccent,
      ),
      Text('Enhanced Counter app!'),
      Icon(
        Icons.add,
        color: Colors.lightGreen,
      ),
    ],
  );
}
}
```

The most important part of the AppTitle widget we have just built is the const keyword in front of the constructor's name. This allows us to create a constant widget to be inserted in the tree or—more specifically—in our Column widget, as follows:

```dart
Column(
  mainAxisSize: MainAxisSize.min,
  children: [
    const AppTitle(),
    // The counter widget
    // The history widget
  ],
),
```

Thanks to constant constructors, we can *cache* widgets and make sure that they're built only *once*. In this specific case, the widget doesn't have external dependencies and it's just a static piece of UI, so it'd be useless if the framework rebuilt it more than once. If you forget to add the `const` keyword in front of your constructor, the linter will warn you!

> **Tip**
> Widgets can have a constant constructor and they can be *cached* to improve your app's performance. Functions, instead, are always rebuilt every time the framework calls the widget's `build` method.

We really discourage you to create functions to return widgets because they *cannot* be marked with the `const` keyword and so they won't be cached. We already looked at the benefits of widgets over functions in the *Understanding the foundations – widgets, elements, and render objects* section. For example, if you decided to not create an `AppTitle` widget, you could have simply used a function, like this:

```
Widget appTitle() {
  return Row(
    mainAxisSize: MainAxisSize.min,
    children: const [
      Icon(Icons.remove),
      Text('Enhanced Counter app!'),
      Icon(Icons.add),
    ],
  );
}
```

It works as intended but you can't write `const appTitle()`, so the widget is rebuilt whenever the framework calls the widget's `build` method.

Let's now move to our app's core: the counter widget.

The counter widget

The second widget we need to build is the counter itself; we're going to place two buttons and set up a state variable to hold the count. In the *Making everything constant* section, we will see that this approach is not really optimal, and we will improve it even more.

However, for now, let's start by converting the `CounterAppBody` widget into a stateful one and adding a `counter` variable, as follows:

```
class _CounterAppBodyState extends State<CounterAppBody> {
  /// The current status of the counter.
  int counter = 0;

  /// Increases the counter by 1.
  void increase() {
    setState(() {
      counter++;
    });
  }

  /// Decreases the counter by 1.
  void decrease() {
    setState(() {
      counter--;
    });
  }

  @override
  Widget build(BuildContext context) {
    return Scaffold(
      body: Center(
        child: Column(
          mainAxisSize: MainAxisSize.min,
          children: [
            const AppTitle(),
            // The counter
            // The history
          ],
        ),
      ),
    );
  }
}
```

We now need to add the two buttons and the counter text. The problem here is that we cannot really extract those into a separated, constant widget as we did with `AppTitle`. The `counter` variable is inside the state itself, so we'd either need to pass a direct dependency, which forbids a `const` constructor, or keep things as they are. For now, let's put everything directly inside the `Column` widget, as follows:

```
Column(
  size: mainAxisSize: MainAxisSize.min,
  children: [
    const AppTitle(),

    Row(
      mainAxisSize: MainAxisSize.min,
      children: [
        ElevatedButton(
          onPressed: decrease,
          child: const Text('-'),
        ),

        Text('$counter'),

        ElevatedButton(
          onPressed: increase,
          child: const Text('+'),
        ),
      ],
    ),
  ],
),
```

At the moment, we cannot move the `Row` widget and its children into a separated widget because the `counter` variable is in this `State` class. We will come back to this in the *Making everything constant* section to see how we can improve the architecture and make more constant widgets.

Since we also want to keep track of the counter value when we press the increment button, we need to add another state variable to remember the value's history, as follows:

```
/// Keeps track of the counter status when '+1' is pressed.
List<int> increaseHistory = [];
```

```
/// Increases the counter by 1.
void increase() {
  setState(() {
    counter++;
    increaseHistory = List<int>.from(increaseHistory)
      ..add(counter);
  });
}
```

Note that we haven't just added a new value to the list, but we have created a new instance. We've done this because we're about to compare this value in HistoryWidget, and in this way, operator== will work as intended. It will compare two different instances, not the same one.

Let's see how the history widget is going to benefit from the comparison of two different list instances.

The history widget

The last UI piece we need to build is a scrolling list underneath the counter. Since we want to keep the build methods fairly short for the sake of readability and separation of UI pieces, we're creating another separated widget called HistoryWidget, like this:

```
/// Keeps track of the counter values whenever it is
/// increased.
class HistoryWidget extends StatefulWidget {
  /// The counters history.
  final List<int> increasesHistory;

  /// Creates an [HistoryWidget] from the given
  /// [increasesHistory].
  const HistoryWidget({
    Key? key,
    required this.increasesHistory,
  }) : super(key: key);
}
class _HistoryWidget extends State<HistoryWidget> {

  @override
  Widget build(BuildContext context) {
```

```dart
  return Column(
    mainAxisSize: MainAxisSize.min,
    children: [
      // The title
      const Text('Increases counter'),

      // The actual list
      Flexible(
        child: Padding(
          padding: const EdgeInsets.symmetric(
            horizontal: 40,
            vertical: 15,
          ),
          child: SizedBox(
            height: 40,
            child: ListView.separated(
              scrollDirection: Axis.horizontal,
              itemCount: widget.increasesHistory.length,
              separatorBuilder: (_, __) {
                return const SizedBox(width: 10);
              }
              itemBuilder: (_, index) {
                return Text(
                  '${widget.increasesHistory[index]}');
              }
            ),
          ),
        ),
      ),
    ],
  );
}
}
}
```

You can see in the previous code block that the list is located outside of this widget, so we need to pass it via a constructor. While this is perfectly fine, it denies the possibility of adding a `const` constructor when instantiating the widget in the tree. We've ended up with this body:

```
Column(
  size: mainAxisSize: MainAxisSize.min,
  children: [
    const AppTitle(),

    // The buttons with the counter
    Row(...),

    HistoryWidget(
      increasesHistory: increasesHistory,
    ),
  ],
),
```

We now have a fully functional app, but we're not entirely satisfied with the result and we think we can do better! Right now, when the widget's `build` method is called, this is the situation:

- The `AppTitle()` widget has a constant constructor and it won't get rebuilt.
- The `Row()` widget isn't constant so its `build` method is executed.
- The `HistoryWidget()` isn't constant either.

When calling `setState(() {})`, the subtree is *always* rebuilt. This means that every time we increase or decrease the counter, we always rebuild the entire contents of `Row` and `HistoryWidget`. We're working on a small app and thus the rebuild cost is negligible, but on larger subtrees, our UI may create a lot of junk. Let's try to fix this!

Manually caching a widget

Let's take a closer look at the functionality of the app to see whether we can save some computational time. Right now, when we increase or decrease the counter, we call `setState(() {})`, and it always rebuilds `HistoryWidget` entirely to correctly show the latest value on the list. The problem is that when we decrease the counter, the history shouldn't be rebuilt because no new entries are added.

In other words, we need to make sure that the contents of `HistoryWidget` are rebuilt only when we tap on +. When tapping on -, we should only decrease the counter and leave the list as it is. Here's how `didUpdateWidget` comes to the rescue:

```
class _HistoryWidgetState extends State<HistoryWidget> {
  /// Manually caching the list.
  late ListView list = buildList();

  /// Building the list.
  ListView buildList() {
    return ListView.separated(…);
  }

  @override
  void didUpdateWidget(covariant HistoryWidget oldWidget) {
    super.didUpdateWidget(oldWidget);

    if (widget.increasesHistory.length !=
        oldWidget.increasesHistory.length) {
      list = buildList();
    }
  }

  @override
  Widget build(BuildContext context) {
    return Column(
      mainAxisSize: MainAxisSize.min,
      children: [
        const Text('Increases counter'),
        SizedBox(
          height: 40,
          child: list,
        ),
      ],
    );
  }
}
```

In the previous code block, you can see that we've refactored HistoryWidget into a stateful widget and stored ListView in a state variable called list. We have also overridden didUpdateWidget to refresh the variable whenever the widget's dependencies change.

Here are a few points to remember:

- We have manually cached the ListView widget so that it will be rebuilt only when required (which is when the list length changes).

- The didUpdateWidget method is called whenever one or more dependencies of the widget change. In our case, we need to make sure that whenever the increasesHistory dependency changes, the cached widget is also updated.

- If we didn't override didUpdateWidget, the list would never be rebuilt again because the list variable would never be assigned again with the new data.

In this particular case, having a function to return a widget is good because the function is *not* inside the build method. The advice we gave in the *Understanding the foundations – widgets, elements, and render objects* section is still valid: try to never use functions inside the build method of a widget. However, when you manually cache a widget using a state variable and didUpdateWidget, it's fine. Make sure to follow this rule of thumb:

- When you use functions *inside* the build method, you cannot control how often the returned child is rebuilt.

- When you use functions *outside* the build method, you can control the returned widget's life cycle and rebuilds thanks to didUpdateWidget, and that is fine.

In general, you should always try to use const constructors, but when you have large subtrees that cannot be constant, you can go for a manual caching strategy! To use another example, Container in Flutter doesn't have a constant constructor, but it may be an immutable piece of the UI. You can see this represented in the following code snippet:

```
Container(
  color: Colors.red
  width: 200,
  height: 160,
  child: VeryBigSubtree(
    color: Colors.red.withAlpha(140),
  ),
),
```

This is an example where manual caching may be useful. This widget may have a fairly big subtree but both `Container` and `VeryBigSubtree` cannot be marked as `const`. Since we know that they are immutable and they don't need to be rebuilt, we can cache them very easily inside the `State` class, like so:

```
final cachedChild = Container(
  color: Colors.red
  width: 200,
  height: 160,
  child: VeryBigSubtree(
    color: Colors.red.withAlpha(140),
  ),
);
```

Since there are no dependencies from widget parameters, we don't even need to override `didUpdateWidget`, and the variable can simply be `final` rather than `late final`. While this approach is quite good, we can still do even better! We can move the state up in the tree and make everything constant.

Making everything constant

At the moment, our app is more efficient than the one we created on the first try, thanks to the manual caching strategy we have just analyzed. However, we can do even better than this by lifting the state up! We're doing all of this in order to remove `setState(() {})` calls and make as many `const` widgets as possible.

We're going to use an `InheritedWidget` to move the state up the tree in order to remove direct dependencies on the state and be able to listen for changes. In this way, we can avoid using `setState(() {})` and have more granular control of rebuilds in our app.

Let's start by creating a model class to hold the counter state. It is going to mix with `ChangeNotifier` because it has the `notifyListener()` method, which can be listened to trigger rebuilds on children. The code is illustrated in the following block:

```
/// The state of the counter in the app.
class Counter with ChangeNotifier {
  int _counter;
  List<int> _history;

  /// Initializes the counter to 0.
  Counter()
```

```
      : _counter = 0,
        _history = [];

  /// The current counter state.
  int get counter => _counter;

  /// The increases history.
  List<int> get history =>
    UnmodifiableListView<int>(_history);

  /// Increases the counter by 1.
  void increase() {
    _counter++;
    _history.add(_counter);

    notifyListeners();
  }

  /// Decreases the counter by 1.
  void decrease() {
    _counter--;
    notifyListeners();
  }
}
```

Since the state is now untied to a specific widget, we can use an `InheritedWidget` to provide the values to any widget down the tree. Any children will be able to access the state directly inside the `build` method without having to ask for values via the constructor.

In the following code block, we're creating an inherited widget that will hold our app's state:

```
class CounterState extends InheritedWidget {
  /// The state of the counter app.
  final Counter model;

  /// Creates a [CounterState] widget.
  const CounterState({
    Key? key,
```

```
    required Widget child,
    required this. model,
  }) : super(key: key, child: child);

  static CounterState of(BuildContext context) {
    return context
        .dependOnInheritedWidgetOfExactType<CounterState>()!;
  }

  @override
  bool updateShouldNotify(covariant CounterState oldWidget) {
    return model.counter != oldWidget. model.counter ||
        model.history.length !=
          oldWidget. model.history.length;
  }
}
```

At this point, our three widgets have no more dependencies from the local widget state, and thus it can also be converted into a Stateless widget. Notice how we're now able to make everything constant without the boilerplate of manual caching, just by using Dart's syntax. We can now make the entire children of the Column constant, as follows:

```
/// The contents of the counter app.
class CounterAppBody extends StatelessWidget {
/// Creates a [CounterAppBody] widget.
  const CounterAppBody({
    Key? key,
  }) : super(key: key);

  @override
  Widget build(BuildContext context) {
    return CounterState(
      model: Counter(),
      child: Scaffold(
        body: Center(
          child: Column(
            mainAxisSize: MainAxisSize.min,
            children: const [
              AppTitle(),
```

```
                    CounterButtons(),
                    HistoryWidget(),
                ],
            ),
          ),
        ),
      );
    }
  }
```

The `CounterButtons` widget simply contains the `Row` widget with the counter and the two increment and decrement buttons. They now alter the state by getting a reference of the inherited widget and directly acting on the model class, which will notify the listeners, as illustrated in the following code snippet:

```
ElevatedButton(
    onPressed: CounterState.of(context).model.decrease,
    child: const Text('-'),
),
```

In `HistoryWidget`, we can now remove `didUpdateWidget` and `buildList` since we don't have a dependency anymore. To listen for updates on the state from the inherited widget, we simply need to wrap the list on an `AnimatedBuilder` widget, and we're done! The `AnimatedBuilder` widget will take care of rebuilding the list only when really needed, leaving the other parts of the widget untouched. The code is illustrated in the following snippet:

```
SizedBox(
    height: 40,
    child: AnimatedBuilder(
        animation: _counterState,
        builder: (context, _) {
            return ListView.separated(
                scrollDirection: Axis.horizontal,
                itemCount: _counterState.history.length,
                itemBuilder: (_, index) {
                    return Text('${_counterState.history[index]}');
                },
                separatorBuilder: (_, __) {
                    return const SizedBox(width: 10);
                }
```

```
      );
    },
  ),
),
```

While it may seem a bit weird, the `AnimatedBuilder` widget is not only used to listen for animations (more on this in *Chapter 7, Building an Animated Excuses Application*) but is also designed to work in conjunction with inherited widgets!

Our counter app is now fully functional and well structured. Thanks to inherited widgets and constant constructors, we've made sure that the framework only rebuilds widgets that need to be updated, leaving the others untouched.

Summary

In this chapter, we learned that Flutter has three trees to efficiently paint the UI. The widget tree is used to configure the underlying `Element` object. The element tree manages the widget life cycle, holds its state (if any), and much more. It's good at comparing because it detects changes and updates or recreates the render object behind. The render object tree is used to paint the UI and handle gestures and does the low-level work.

With this information, we have built a simple but efficient counter app that keeps a small history of increased values. We have learned about the importance of constant constructors and saw a very effective technique to manually cache widgets. We could say that the more constant constructors we can use, the better.

In the last section, the app has been improved even more with the usage of an inherited widget that lifted the state up and allowed for more constant constructors.

The next chapter is about techniques to build UIs in Flutter. We will be building a race standings app using Flutter's material library to show how its layout widgets are best used.

Further reading

- `InheritedWidget`: https://api.flutter.dev/flutter/widgets/ InheritedWidget-class.html

- `setState`: https://api.flutter.dev/flutter/widgets/State/ setState.html

- `Element`: https://api.flutter.dev/flutter/widgets/Element-class.html

- `RenderObject`: https://api.flutter.dev/flutter/rendering/ RenderObject-class.html

2
Building a Race Standings App

In this chapter, we'll create a more complex project than the counter app we built previously. We're going to create the UI of a generic racing game that shows both the results of races and drivers' standings.

Other than code quality, we will also pay a lot of attention to the **user experience** (**UX**); this includes localization, internationalization, responsiveness, color contrast, and more, which will be used to create a high-quality result. We will also learn how to manually work with the device's pixels with `CustomPainter`, for those cases where Flutter widgets aren't enough.

In this chapter, we will cover the following topics:

- Creating responsive screens using the `LayoutBuilder` widget
- Using the `intl` package to localize the app
- Working with images – PNGs and vectorial files
- Using custom painters to paint complex UI elements

Let's get started!

Technical requirements

We recommend that you work on the `stable` channel and work with Flutter version 2.5 or newer. Any version after Flutter 2.0 could still be okay, but we can't guarantee that you won't encounter unexpected problems while trying to compile our source code.

Since we're going to test the UI on various screen sizes, we will compile it for the web so that we can resize the browser window to easily emulate different viewports. While very convenient and quick to test, you could spin up various emulators with different screen sizes or use your own physical devices.

We also recommend using either Android Studio or **Visual Studio Code (VS Code)**: choose the one you like more!

The complete source code for this project can be found at `https://github.com/ PacktPublishing/Cross-Platform-UIs-with-Flutter/tree/main/chapter_2`.

Setting up the project

Before we start creating the app, we need to prepare the environment and make sure we care about the UX from the beginning.

Create a new Flutter project in your favorite IDE and make sure that you enable web support by clicking on the **Add Flutter web support** checkbox. A basic `analysis_options.yaml` file will be created for you. Even if it's not strictly required, we strongly recommend that you add more rules to enhance your overall code quality.

> **Tip**
>
> If you want to easily set up the `analysis_options.yaml` file with the configuration we've recommended, just go to this project's GitHub repository and copy the file into your project! You can also find a quick overview of the rules in the *Setting up the project* section of *Chapter 1, Building a Counter App with History Tracking to Establish Fundamentals*.

Since we aren't uploading this project to `https://pub.dev/`, make sure that your `pubspec. yaml` file has the `publish_to: 'none'` directive uncommented. Before we start coding, we still need to set up localization, internationalization, route management, and custom text fonts.

Localization and internationalization

Localizing an app means adapting the content according to the device's geographic settings to appeal to as many users as possible. In practical terms, for example, this means that an Italian user and an American user would see the same date but in different formats. In Italy, the date format is `d-m-y`, while in America, it's `m-d-y`, so the app should produce different strings according to the device's locale. It's not only about the date, though, because localizing can also mean the following changes occur:

- Showing prices with the proper currency (Euro, Dollar, Sterling, and so on)
- Taking into account time zones and offsets while displaying dates for events
- Choosing between a 24-hour or 12-hour time format
- Deciding which decimal separator is used (a comma, a full stop, an apostrophe, and so on)

Internationalizing, which is part of the localization process, means translating your app's text according to the device's locale. For example, while an Italian user would read *Ciao!*, an American user would read *Hello!*, and all of this is done automatically by the app.

Setting up localization support in Flutter is very easy! Start by adding the SDK direct dependency to the `pubspec.yaml` file:

```
dependencies:
  flutter:
    sdk: flutter

  flutter_localizations:
    sdk: flutter

  intl: ^0.17.0
```

The `intl` package is maintained by the Dart team and offers numerous internationalization and localization utilities we will explore throughout this chapter, such as `AppLocalization` and `DateFormat`.

Still in the `pubspec` file, we need to add another line at the bottom of the `flutter` section:

```
flutter:
  generate: true
```

We must do this to bundle the various localization files into our app so that the framework will be able to pick the correct one based on the user's locale settings.

The last file we need to create must be located at the root of our project, it must be called `l10n.yaml` *exactly*, and it must have the following contents:

```
arb-dir: lib/localization/l10n
template-arb-file: app_en.arb
output-localization-file: app_localizations.dart
```

This file is used to tell Flutter where the translations for the strings are located so that, at runtime, it can pick up the correct files based on the locale. We will be using English as the default, so we will set `app_en.arb` to `template-arb-file`.

Now, we need to create two ARB files inside `lib/localization/l10n/` that will contain all of our app strings. The first file, `app_en.arb`, internationalizes our app in English:

```
{
    "app_title": "Results and standings",
    "results": "Results",
    "standings": "Standings",
}
```

The second file, `app_it.arb`, internationalizes our app in Italian:

```
{
    "app_title": "Risultati e classifiche",
    "results": "Risultati",
    "standings": "Classifica",
}
```

Every time you add a new entry to the ARB file, you must make sure that the ARB keys match! When you run your app, automatic code generation will convert the ARB files into actual Dart classes and generate the `AppLocalization` class for you, which is your reference for translated strings. Let's look at an example:

```
final res = AppLocalizations.of(this)!.results;
```

If you're running your app on an Italian device, the value of the `res` variable will be `Risultati`. On any other device, the variable would hold `Results` instead. Since English is the default language, when Flutter cannot find an ARB file that matches the current locale, it will fall back to the default language file.

> **Tip**
> Whenever you add or remove strings to/from your ARB files, make sure you always hit the **Run** button of your IDE to build the app! By doing this, the framework builds the newly added strings and bundles them into the final executable.

Extension methods, which are available from Dart 2.7 onwards, are a very nice way to add functionalities to a class without using inheritance. They're generally used when you need to add some functions or getters to a class and make them available for any instance of that type.

Let's create a very convenient `extension` method to be called directly on any Flutter string, which reduces the boilerplate code and saves some `import` statements. This is a very convenient shortcut to easily access the localization and internationalization classes that's generated by Flutter. The following is the content of the `lib/localization/localization.dart` file:

```
import 'package:flutter/material.dart';
import 'package:flutter_gen/gen_l10n/app_localizations.dart';
export 'package:flutter_gen/gen_l10n/app_localizations.dart';


/// Extension method on [BuildContext] which gives a quick
/// access to the `AppLocalization` type.
extension LocalizationContext on BuildContext {
  /// Returns the [AppLocalizations] instance.
  AppLocalizations get l10n => AppLocalizations.of(this)!;
}
```

With this code, we can simply call `context.l10n.results` to retrieve the internationalized value of the *Results* word.

Last, but not least, we need to make sure that our root widget installs the various localization settings we've created so far:

```
/// The root widget of the app.
class RaceStandingsApp extends StatelessWidget {
  /// Creates an [RaceStandingsApp] instance.
  const RaceStandingsApp ({Key? key}) : super(key: key);

  @override
  Widget build(BuildContext context) {
    return MaterialApp(
      // Localized app title
      onGenerateTitle: (context) => context.l10n.app_title,

      // Localization setup
      localizationsDelegates:
        AppLocalizations.localizationsDelegates,
      supportedLocales: AppLocalizations.supportedLocales,
```

```
    // Routing setup
    onGenerateRoute: RouteGenerator.generateRoute,

    // Hiding the debug banner
    debugShowCheckedModeBanner: false,
  );
}
}
```

If you're dealing with multiple languages and a lot of strings, manually working on ARB files may become very hard and error-prone. We suggest that you either look at Localizely, an online tool that handles ARB files, or install an ARB plugin manager in your IDE.

In the preceding code, you may have noticed the RouteGenerator class, which is responsible for route management. That's what we're going to set up now!

Routes management

In this app, we're using Flutter's built-in routing management system: the Navigator API. For simplicity, we will focus on Navigator 1.0; in *Chapter 5, Exploring Navigation and Routing with a Hacker News Clone*, we will cover the routing topic in more depth.

Let's create a simple RouteGenerator class to handle all of the routing configurations of the app:

```
abstract class RouteGenerator {
  static const home = '/';

  static const nextRacesPage = '/next_races';

  /// Making the constructor private since this class is
  /// not meant to be instantiated.
  const RouteGenerator._();

  static Route<dynamic> generateRoute(RouteSettings
    settings) {
    switch (settings.name) {
      case home:
        return PageRouteBuilder<HomePage>(
          pageBuilder: (_, __, ___) => const HomePage(),
        );
```

```
    case nextRacesPage:
      return PageRouteBuilder<NextRacesPage>(
        pageBuilder: (_, __, ___) =>
          const NextRacesPage(),
      );

    default:
      throw const RouteException('Route not found');
    }
  }
}

/// Exception to be thrown when the route doesn't exist.
class RouteException implements Exception {
  final String message;

  /// Requires the error [message] for when the route is
  /// not found.
  const RouteException(this.message);
}
```

By doing this, we can use `Navigator.of(context).pushNamed(RouteGenerator.home)` to navigate to the desired page. Instead of hard-coding the route names or directly injecting the PageRouteBuilders in the methods, we can gather everything in a single class.

As you can see, the setup is very easy because all of the management is done in Flutter internally. We just need to make sure that you assign a name to the page; then, `Navigator` will take care of everything else.

Last but not least, let's learn how to change our app's look a bit more with some custom fonts.

Adding a custom font

Instead of using the default text font, we may wish to use a custom one to make the app's text look different from usual. For this purpose, we need to reference the `google_fonts` package in the dependencies section of `pubspec` and install it:

```
return MaterialApp(
  // other properties here...
```

```
theme: ThemeData.light().copyWith(
  textTheme: GoogleFonts.latoTextTheme(),
),
);
```

It couldn't be easier! Here, the Google font package fetches the font assets via HTTP on the first startup and caches them in the app's storage. However, if you were to manually provide font files assets, the package would prioritize those over HTTP fetching. We recommend doing the following:

1. Go to `https://fonts.google.com` and download the font files you're going to use.

2. Create a top-level directory (if one doesn't already exist) called `assets` and then another sub-folder called `fonts`.

3. Put your font files into `assets/fonts` without renaming them.

4. Make sure that you also include the `OFL.txt` license file, which will be loaded at startup. Various license files are included in the archive you downloaded from Google Fonts.

Once you've done that, in the `pubspec` file, make sure you have declared the path to the font files:

```
assets:
  - assets/fonts/
```

The package only downloads font files if you haven't provided them as assets. This is good for development but for production, it's better to bundle the fonts into your app executable to avoid making HTTP calls at startup. In addition, font fetching assumes that you have an active internet connection, so it may not always be available, especially on mobile devices.

Finally, we need to load the license file into the licenses registry:

```
void main() {
  // Registering fonts licences
  LicenseRegistry.addLicense(() async* {
    final license = await rootBundle.loadString(
      'google_fonts/OFL.txt',
    );

    yield LicenseEntryWithLineBreaks(['google_fonts'],
      license);
  });

  // Running the app
```

```
  runApp(
    const RaceStandingsApp(),
  );
}
```

Make sure that you add the `LicenseRegistry` entry so that, if you use the `LicensePage` widget, the licensing information about the font is bundled into the executable correctly.

Now that we've set everything up, we can start creating the app!

Creating the race standings app

The home page of the app is going to immediately provide our users with a quick way to see both race results and driver standings. Other kinds of information, such as upcoming races or a potential settings page, should be placed on other pages.

This is what the app looks like on the web, desktop, or any other device with a large horizontal viewport:

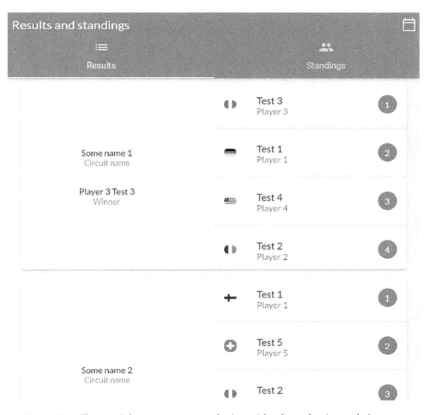

Figure 2.1 – The app's home page on a device with a large horizontal viewport

Having everything on a single screen would make the UI too *dense* because there would be too much information for the user to see. Tabs are great when it comes to splitting contents into multiple pages and they're also very easy to handle – it's just a matter of swiping!

On mobile devices, or smaller screen sizes, the home page looks like this:

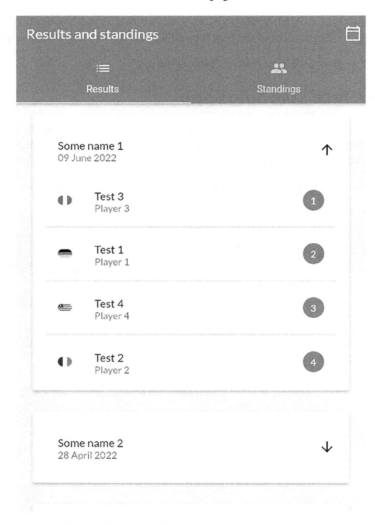

Figure 2.2 – The app's home page on smaller devices

As you can see, since there is less horizontal space, we need to rearrange the contents so that it fits with less space. Laying down contents on two columns would take too much space, so we've decided to create a sort of dropdown menu. The black arrow slides up and down to show or hide contents. The app has two main pages:

- The HomePage widget, where we show the results of the past races and the current drivers' standings.
- The NextRaces widget, where we show a brief list of the upcoming races.

Now, let's start creating the HomePage widget!

The HomePage widget

The home page is going to have two main tabs to display core information. This already gives us a pretty important hint regarding what we need to do: we need to create two widgets to hold the contents of each tab and we want them to be constant.

The following is the build() method for the HomePage widget:

```
@override
  Widget build(BuildContext context) {
    return DefaultTabController(
      length: 2,
      child: Scaffold(
        appBar: AppBar(
          title: Text(context.l10n.app_title),
          elevation: 5,
          bottom: TabBar(
            tabs: [
              Tab(
                icon: const Icon(Icons.list),
                text: context.l10n.results,
              ),
              Tab(
                icon: const Icon(Icons.group),
                text: context.l10n.standings,
              ),
            ],
          ),
```

```
        ),
      body: const TabBarView(
        children: [
          ResultsTab(),
          StandingsTab(),
        ],
      ),
    ),
  );
}
```

Thanks to widget composition, we can use `const TabBarView` because both children have a constant constructor. Now, let's learn how to build the `ResultsTab` and `StandingsTab` widgets.

The results tab

This page is responsive because it dynamically rearranges its contents to best fit the current horizontal and vertical viewport constraints. In other words, this widget lays out the contents in different ways based on the different screen sizes, thanks to `LayoutBuilder`:

```
return LayoutBuilder(
  builder: (context, dimensions) {
    // Small devices
    if (dimensions.maxWidth <= mobileResultsBreakpoint) {
      return ListView.builder(
        itemCount: resultsList.length,
        itemBuilder: (context, index) =>
          _CompactResultCard(
          results: resultsList[index],
        ),
      );
    }

    // Larger devices
    return Padding(
      padding: const EdgeInsets.symmetric(
        vertical: 20,
      ),
```

```
        child: ListView.builder(
          itemCount: resultsList.length,
          itemBuilder: (context, index) =>
            _ExpandedResultCard(
            results: resultsList[index],
          ),
        ),
      );
    },
  );
```

Here, the `mobileResultsBreakpoint` constant has been put in `lib/utils/breakpoints.dart`. We are gathering all of our responsive breakpoint constants into a single file to simplify both maintenance and testing. Thanks to `LayoutBuilder`, we can retrieve the viewport dimensions and decide which widget we want to return.

The `ExpandedResultCard` widget is meant to be displayed or larger screens, so we can safely assume that there is enough horizontal space to lay down contents in two columns. Let's learn how to do this:

```
Card(
  elevation: 5,
  child: Row(
    children: [
      // Race details
      Expanded(
        flex: leftFlex,
        child: Column(
          mainAxisSize: MainAxisSize.min,
          children: [ … ],
        ),
      ),

      // Drivers final positions
      Expanded(
        flex: 3,
        child: DriversList(
          results: results,
```

```
        ),
      ),
    ],
  ),
),
```

To make this widget even more responsive, we can also control the relative widths of the columns. We're still using LayoutBuilder to decide on the flex of the Expanded widget to ensure that the content fits the space in the best possible way:

```
return LayoutBuilder(
  builder: (context, dimensions) {
    var cardWidth = max<double>(
      mobileResultsBreakpoint,
      dimensions.maxWidth,
    );

    if (cardWidth >= maxStretchResultCards - 50) {
      cardWidth = maxStretchResultCards;
    }

    final leftFlex =
      cardWidth < maxStretchResultCards ? 2 : 3;

    return Center(
      child: SizedBox(
        width: cardWidth - 50,
        child: Card( ... ),
      ),
    );
  },
);
```

Here, we compute the overall width of the surrounding Card and then determine the widths of the columns by computing the flex value.

The _CompactResultCard widget is meant to be displayed on smaller screens, so we need to arrange the widgets along the vertical axis using a single column. To do this, we must create a simple widget called Collapsible that has a short header and holds the contents on a body that slides up and down:

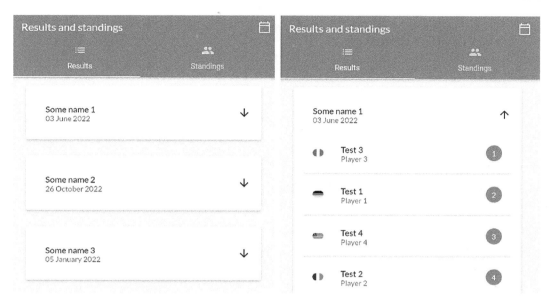

Figure 2.3 – On the left, the content is hidden; on the right, the content is visible

This approach is very visually effective because, considering there isn't much horizontal space available, we immediately show the most important information. Then, if the user wants to know more, they can tap the arrow to reveal additional (but still important) information. First, we store the open/closed state of the card in an inherited widget:

```
/// Handles the state of a [Collapsible] widget.
class CollapsibleState extends InheritedWidget {
  /// The state of the [Collapsible] widget.
  final ValueNotifier<bool> state;

  /// Creates a [CollapsibleState] inherited widget.
  const CollapsibleState({
    Key? key,
    required this.state,
    required Widget child,
  }) : super(key: key, child: child);

  /// Conventional static access of the instance above the
  /// tree.
  static CollapsibleState of(BuildContext context) {
```

```
      return context.dependOnInheritedWidgetOfExactType<
        CollapsibleState>()!;
    }


    @override
    bool updateShouldNotify(CollapsibleState oldWidget) =>
        state != oldWidget.state;
  }
```

Then, we use the `SizeTransition` widget to make the contents underneath appear and disappear with a sliding transition. The animation is driven by `ValueListenableBuilder`:

```
  return ValueListenableBuilder<bool>(
    valueListenable: CollapsibleState.of(context).state,
    builder: (context, value, child) {
      if (!value) {
        controller.reverse();
      } else {
        controller.forward();
      }

      return child!;
    },
    child: Padding(
      padding: widget.edgeInsets,
      child: Column(
        mainAxisSize: MainAxisSize.min,
        crossAxisAlignment: CrossAxisAlignment.start,
        children: regions,
      ),
    ),
  );
```

We use the `child` parameter to ensure that the builder won't unnecessarily rebuild `Column` over and over. We only need to make sure that `reverse()` or `forward()` is called whenever the boolean's state is changed.

Since both `_CompactResultCard` and `_ExpandedResultCard` need to display a date, we have created a `mixin` for the state class to be able to easily share a common formatting method:

```
/// A date generator utility.
mixin RandomDateGenerator on Widget {
  /// Creates a random date in 2022 and formats it as 'dd
  /// MMMM y'.
  /// For more info on the format, check the [Intl]
  /// package.
  String get randomDate {
    final random = Random();

    final month = random.nextInt(12) + 1;
    final day = random.nextInt(27) + 1;

    return DateFormat('dd MMMM y').format(
      DateTime(2022, month, day + 1),
    );
  }
}
```

The `DateFormat` class is included in the `intl` package, and it can automatically translate the date string into various languages. In this case, the `'dd MMMM y'` combination prints the day in double digits, the name of the month with a capital letter, and the year in 4-digit format.

Tip

You can format the date in many ways – you just need to change the tokens in the string. We won't cover them all here because there are thousands of possible combinations; if you do want to know more, we recommend that you look at the documentation: `https://pub.dev/documentation/intl/latest/intl/DateFormat-class.html`.

Now, let's create the drivers' standings tab.

The drivers' standings tab

Even though this page contains a simple list of people and their country flags and scores, there are still some considerations to make. The first one is that we don't want to always use the entirety of the viewport's width, like this:

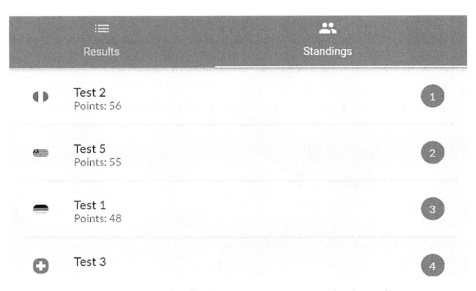

Figure 2.4 – Example of bad space management in the drivers list

The user may have trouble gathering all of the information at first glance because there is too much space between the important UI parts. We need to make sure that the content can shrink to fit smaller sizes, but we don't want to always use the entire available width.

As shown in the preceding screenshot, always using the entire horizontal viewport may lead to bad user experiences. To avoid this, we're going to set up a breakpoint that limits how the list can grow in the horizontal axis:

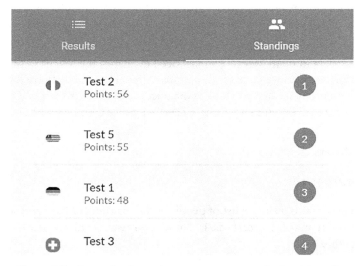

Figure 2.5 – Example of good space management in the drivers list

Here, we've created a new breakpoint called maxStretchStandingsCards that imposes horizontal bounds to the list so that it doesn't grow too much. This is how the standings list is being built:

```
ListView.separated(
  shrinkWrap: true,
  itemCount: standingsList.length,
  itemBuilder: (context, index) {
    final item = standingsList[index];

    return ListTile(
      title: Text(item.name),
      subtitle: Text('${context.l10n.points}:
        ${item.points}'),
      leading: Column( ... ),
      trailing: NumberIndicator( ... ),
    );
  },
  separatorBuilder: (_, __) {
    return const Divider(
      thickness: 1,
      height: 10,
    );
  }
),
```

The official Flutter documentation states that both ListView.builder() and ListView.separated() are very efficient builders when you have a fixed, long list of children to paint. They build children on demand because the builder is only called on visible widgets.

We could have achieved the same result by wrapping a Column in a scrollable widget, but it wouldn't be as efficient as using lazy builders, as we did in the previous code block. For example, we *don't* suggest that you do this with fixed-length lists:

```
SingleChildScrollView(
  child: Column(
    children: [
      for (item in itemsList)
        item,
    ],
```

```
    ),
  )
```

The `Column` widget always renders all of its children, even if they're out of the currently visible viewport. If the user doesn't scroll the column, the widgets that aren't in the viewport would still be rendered, even if they never appeared on the screen. This is why we suggest that you use list builders rather than columns when you have a long list of widgets to render.

Another point we want to touch on is using SVG and PNG files for images. We've been using both and we recommend that you do too because vectorial images are not always a good choice.

Vectorial images guarantee that you keep the quality high on scaling, and probably have a smaller file size than a PNG, but they may be very complicated to parse. PNGs may not scale very well but they're quick to load and, when compressed, they can be really small. Here are some suggestions:

- Always compress the SVG and PNG files you use to make sure they occupy the least possible amount of memory.

- When you see that the SVG file is big and takes a few seconds to load, consider using a PNG image instead.

- When you know that the image is going to scale a lot and the width/height ratio may now linearly change, consider using vectorial images instead.

In this project, we have used PNG images for country flags since they're small, and we aren't resizing them.

For our vectorial assets, we've used a popular and well-tested package called `flutter_svg` that makes managing vectorial assets very easy. For example, here's how we load an SVG file in the project:

```
SvgPicture.asset(
  'assets/svg/trophy.svg',
  width: square / 1.8,
  height: square / 1.8,
  placeholderBuilder: (_) => const Center(
    child: CircularProgressIndicator(),
  ),
),
```

We can dynamically define its dimensions with `width` and `height` and also use `placeholderBuilder` to show a progress indicator in case the file vectorial was expensive to parse.

Now, let's create the `NextRaces` widget.

The NextRaces widget

While showing the upcoming races of the championship is still part of the app, this isn't its primary focus. The user can still check this data but it's optional, so let's create a new route to *hide* it on a separated page. So that we don't have a static list with a few colors on it, we want to split the page into two main pieces:

- At the top, we want to show how many races there are left in the championship. To make it visually attractive, we've used an image and a fancy circular progress indicator.

- At the bottom, we have the list of upcoming races.

The page is simple, but it only shows data about the upcoming races and nothing more. We haven't filled the UI with distracting background animations, low-contrast colors, or widgets that are too complex.

> **Tip**
> Always try to strive for a good balance between providing the necessary content and making the app as simple as possible. Having too many animations, images, or content on a page might be distracting. However, at the same time, a UI that is too minimal may not impress the user and give the feeling of a poorly designed app.

Here's what the **Next races** UI is going to look like:

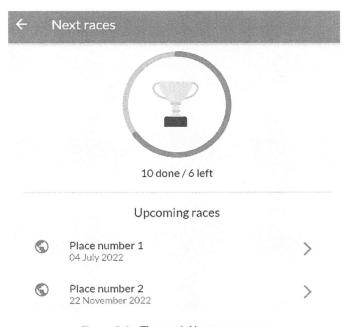

Figure 2.6 – The app's Next races page

At the top, you can see a trophy surrounded by something similar to `CircularProgressIndicator`. Flutter doesn't have a widget that allows us to achieve that exact result and nor do we have an easy way to build it. We may start with a `Stack` but then the rail and the progress bar may be difficult to render with common widgets.

In this case, we want to create a specific widget with particular constraints and shapes that's not built in the Flutter framework. All of these hints lead us in a single direction: custom painters! Once again, we're making the sizes responsive by dynamically calculating the width and height using the `square` variable:

```
LayoutBuilder(
  builder: (context, dimensions) {
    final square = min<double>(
      maxCircularProgress,
      dimensions.maxWidth,
    );

    return Center(
      child: CustomPaint(
        painter: const CircularProgressPainter(
          progression: 0.65,
        ),
        child: SizedBox(
          width: square,
          height: square,
          child: Center(
            child: SvgPicture.asset(
              'assets/svg/trophy.svg',
            ),
          ),
        ),
      ),
    );
  }
);
```

Thanks to `CustomPaint`, we can normally render a child and *additionally* paint some custom graphics in the background using the `painter` parameter. In the same way, we could have painted the same circular progress indicator in the foreground using `foregroundPainter`.

Custom painters aren't the easiest thing to use but they give you a lot of power. You're given a `Canvas` object where you can paint everything: lines, Bézier curves, shapes, images, and more. Here's how we've created the painter for the circular progress indicator:

```
/// A circular progress indicator with a grey rail and a
/// blue line.
class CircularProgressPainter extends CustomPainter {
  /// The progression status.
  final double progression;

  /// Creates a [CircularProgressPainter] painter.
  const CircularProgressPainter({
    required this.progression,
  });

  @override
  void paint(Canvas canvas, Size size) {
    // painting the arcs...
  }

  @override
  bool shouldRepaint(covariant CircularProgressPainter old)
  {
    return progression != old.progression;
  }
}
```

We need to extend `CustomPainter` and override two very important methods:

* `shouldRepaint`: This method tells the custom painter when it should repaint the contents. If you have no external dependencies, this method can safely just `return false`. In our case, if the progression changes, we need to also change the arc span, so we need to check whether `progression != old.progression`.

* `paint`: This method provides a `Canvas`, along with its dimensions. It's responsible for painting the content to the UI.

Here's how we have implemented `paint` to draw the arcs:

```
// The background rail
final railPaint = Paint()
  ..color = Colors.grey.withAlpha(150)
  ..strokeCap = StrokeCap.round
  ..style = PaintingStyle.stroke
  ..strokeWidth = 8;

// The arc itself
final arcPaint = Paint()
  ..color = Colors.blue
  ..strokeCap = StrokeCap.round
  ..style = PaintingStyle.stroke
  ..strokeWidth = 8;

// Drawing the rail
final center = size.width / 2;
canvas.drawArc(
  Rect.fromCircle(
    center: Offset(center, center),
    radius: center,
  ),
  -pi / 2,
  pi * 2,
  false,
  railPaint,
);

// Drawing the arc
canvas.drawArc(
  Rect.fromCircle(
    center: Offset(center, center),
    radius: center,
  ),
  -pi / 2,
  pi * 2 * progression,
```

```
    false,
    arcPaint,
);
```

The `Paint` class defines the properties (thickness, color, border fill style, and more) of the lines or shapes we're going to paint, while the `Canvas` class contains a series of methods for drawing various things on the UI, such as the following:

- `drawLine`
- `drawCircle`
- `drawImage`
- `drawOval`
- `drawRect`
- `clipPath`

And much more! Some mathematical skills are required here because we need to compute the arc length of the progress bar based on the progression percentage. The background track is just a full arc, so it's easy to paint. On the other hand, the swipe of the progress bar needs to start from the top (`-pi / 2`) and be as wide as the percentage allows (`pi * 2 * progression`).

We've done it! The app now has two main pages: the first one shows rankings and standings, while the other one is about the upcoming races in the championship.

Summary

In this chapter, we learned how internationalization and localization work in Flutter and we also used some custom fonts from Google Fonts. Thanks to the `intl` package, we can, for example, format currencies and dates based on the device's locale.

The race standings app is responsive because it dynamically rearranges the UI elements based on the viewport's sizes. Thanks to breakpoints and the `LayoutBuilder` widget, we were able to easily handle the screen size changes.

The `builder()` and `separated()` constructors of ListViews are very efficient when it comes to painting a fixed series of widgets since they lazily load children.

We also used both PNG and SVG image assets. To render more complex widgets, such as the circular progress indicator, we used `CustomPainter` to go a bit more *low level*.

In the next chapter, we're going to cover Flutter's built-in state management solution: `InheritedWidget`. We will also use the popular `provider` package, which is a wrapper of `InheritedWidget` that's easier to use and test.

Further reading

For more information about the topics that were covered in this chapter, take a look at the following resources:

- Localization in Flutter: `https://docs.flutter.dev/development/accessibility-and-localization/internationalization`
- Assets and images: `https://docs.flutter.dev/development/ui/assets-and-images`
- The `flutter_svg` package: `https://pub.dev/packages/flutter_svg`
- The `intl` package: `https://pub.dev/packages/intl`

3

Building a Todo Application Using Inherited Widgets and Provider

Quite often, when building an application, developers must also figure out how to share information in multiple places in the application. A common pattern for sharing data is to pass information down as a property top-down from parent to child, and through each nested child until you reach the widget that depends on that information. While this pattern is certainly fine for small applications, it can become quite cumbersome in large applications. In this chapter, we'll explore this pattern and alternative solutions for sharing dependencies in an application.

In this chapter, we will cover the following topics:

- Sharing dependencies in a Flutter application
- Setting up the project
- Creating the Todo application

Let's begin!

Technical requirements

Make sure that your Flutter environment has been updated to the latest version in the `stable` channel. Clone our repository and use your favorite IDE to open the Flutter project we've built at `chapter_3/start`.

The project that we'll build upon in this chapter can be found on GitHub: `https://github.com/PacktPublishing/Flutter-UI-Projects-for-iOS-Android-and-Web/tree/main/chapter_3/start`.

The complete source code can be found on GitHub as well: `https://github.com/PacktPublishing/Flutter-UI-Projects-for-iOS-Android-and-Web/tree/main/chapter_3/end`.

Sharing dependencies in a Flutter application

As mentioned previously, one common pattern for sharing information in a Flutter application is to pass information down as a property from the top down from parent to child, and through each nested child until you reach the widget that depends on that information. We will refer to this process as **property forwarding**.

While this strategy is certainly fine in limited instances, it is not ideal when sharing information in several components or several nested tree layers. It is also not efficient because if it's not handled properly, a state change in one of the properties can potentially trigger a rebuild of large portions of the widget tree.

Let's look at the following diagram to get a better understanding of this problem:

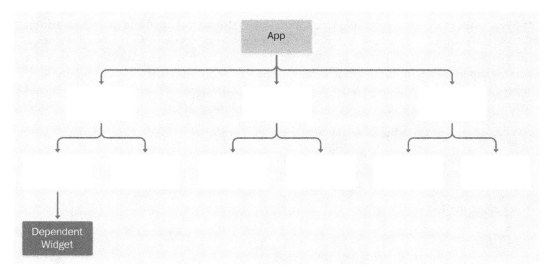

Figure 3.1 – A Flutter application widget tree

Here, we have several nested layers of widgets. At the top level, our **App** widget contains some business data that a descendant three levels below cares about. Using property forwarding, we can propagate that data down the widget tree through each descendant until the data has reached the **Dependent Widget**.

Let's review how this would look:

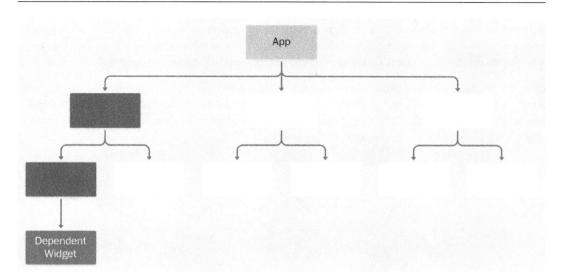

Figure 3.2 – A Flutter application widget tree with property forwarding

The preceding diagram shows the data that has been defined in the **App** widget is forwarded down to the widget that depends on it and that we have solved our problem, albeit in a very cumbersome way. But what happens if another widget has the same dependency? Take a look at the following diagram:

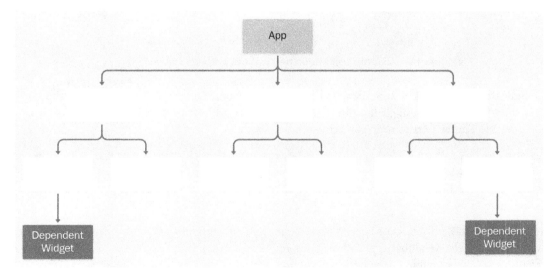

Figure 3.3 – A Flutter application tree with property forwarding with multiple dependent widgets

The preceding diagram shows a widget tree with multiple dependencies in deeply nested child widgets. Now, we must pass the application data through two completely independent trees. The effort that we put forth to maintain dependencies using property forwarding will continue to grow.

Ideally, we would prefer to define data in our App widget and reference it without forwarding the information down multiple layers of uninterested widgets. The following diagram represents our ideal solution, where we can define a **Data Provider** in our **App** widget and reference that **Data Provider** further down the tree in our **Dependent Widget**:

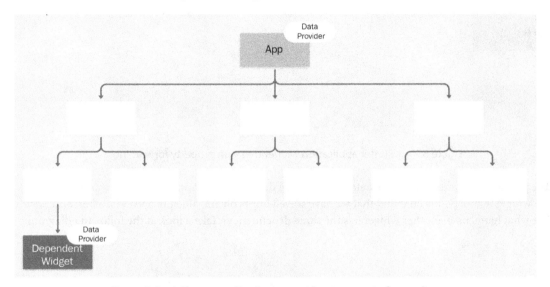

Figure 3.4 – A Flutter application tree without property forwarding

To handle the propagation of dependencies in a less cumbersome and more efficient way, we should define these properties in a single location and reference them without property forwarding. For this, we can turn to Flutter's InheritedWidget base class.

An InheritedWidget allows information to be propagated down Flutter's widget tree. Put another way, this special widget allows descendants in the tree to access its values without using property passing, while also allowing those descendants to subscribe to changes to the InheritedWidget, causing them to be rebuilt.

The InheritedWidget class is used throughout the Flutter framework. If you have referenced Theme.of to retrieve global application styles, or MediaQuery.of to retrieve the size of the current media (for example, the window containing your app), then you have already experienced some of the capabilities of InheritedWidget.

An InheritedWidget is not the same as state management, but it is often used in conjunction with other types of classes, such as ChangeNotifier and ValueNotifier, which allow you to manage application state and business logic.

Think of InheritedWidget as a tool for dependency injection rather than state management: *a technique in which an object receives other objects that it depends on from another object.*

Now that we understand the value of InheritedWidget in providing dependencies, let's set up our project.

Setting up the project

In this chapter, we will learn how to use inherited widgets by refactoring a simple Todo application. The application has an input field to create new tasks, or todos, and a list that will display todos and allow a user to either complete them or delete them.

Here is how our example should look when it is running:

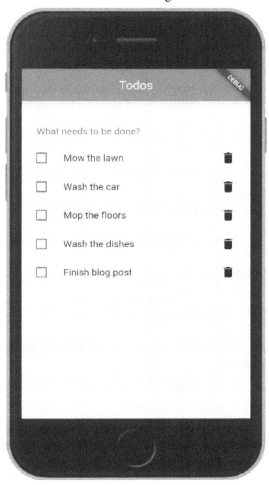

Figure 3.5 – A simple Todo application

We will start with an example that shares dependencies by forwarding application data down the widget tree as properties. We will change this example to propagate those same dependencies using `InheritedWidget` rather than using property forwarding. Then, we will refactor that same example to use the `Provider` package, which is endorsed by the Flutter community and Google as a pragmatic solution to dependency injection and state management.

After downloading the initial application from GitHub, we can start by moving into its root directory and running the `flutter pub get` command from the terminal. This will install any package dependencies that the application needs. Additionally, opening the project in an IDE will also cause the project to be installed when it's loaded and when the `pubspec.yaml` file is updated.

This project uses Flutter's new skeleton template, which is the Flutter team's take on the best practices and structure of a complete application. This template can be used to generate new projects by executing `flutter create -t skeleton` in your terminal, but we will start from a baseline that already includes a `ChangeNotifier`.

After installing the application's dependencies, execute the `flutter run` command from the same terminal window and view the project either in your browser or on your currently running emulator.

Because this application uses a template that has a different folder structure, it may be much more advanced than the versions in previous chapters. Here is the folder structure:

```
lib
|-- src:
    |-- controllers: Controllers glue data to Flutter
        Widgets.  Controllers defined at this level will be
        used in multiple features, or globally throughout
        the application.
    |-- todos_vanilla
    |-- widgets
|-- main.dart
```

Let's briefly explore the purpose of each folder:

- `src`: The source folder contains implementation details of our application that, unless exposed, will be hidden from consumers.
- `controllers`: Controllers glue data to Flutter widgets. Controllers defined at this level will be used in multiple features, or globally throughout the application.
- `todos_vanilla`: A feature folder, normally defined as a top-level folder that represents major parts of the application. Features can include their own controllers, models, services, and UI and can import from shared folders but not other features.

- `widgets`: Widgets defined at this level will be used in multiple features, or globally throughout the application.

- `main.dart`: The entry point of the application.

One Flutter class that we have not touched on so far is `ChangeNotifier`, which we are using as a mixin for our `TodoController`. A `ChangeNotifier`, which inherits from `Listenable`, is a class that can be used to tell parts of the application that information has been changed. It operates very similarly to `StatefulWidget`, with a `notifyListeners` function that can be used to trigger updates, much like `setState`.

In the next section, we will examine the starter application and refactor it so that it uses `InheritedWidget` instead of property forwarding.

Creating the Todo application

In this section, we'll build a Todo application that relies on an `InheritedWidget` – specifically, `InheritedNotifier` – to share and modify a list of todos using a `ChangeNotifier`. You will be able to create a Todo, mark it as completed, or delete it.

The starting point of this section is an application that already contains `TodoController`, which we will be sharing with our `InheritedNotifier`. Once we have dealt with `InheritedWidget` and have a clear understanding of how everything works, we will move on to the `Provider` exercise.

Using InheritedWidget for state management

Before we begin refactoring our starting application to use `InheritedWidget` to share data, let's get familiar with the process of sharing information by forwarding properties down the widget tree.

Starting with property passing

Open the `main.dart` file; you will discover that we are creating our `TodosController` and passing it into our `TodosApp`:

```
void main() async {
  final todoController = TodosController();
  runApp(TodosApp(todoController: todoController));
}
```

In this code block, we run the application and pass `TodosController`. The app will listen to `TodosController` for changes.

Next, navigate to TodosApp; in the onGenerateRoute function of MaterialApp, you will notice that we are passing the controller into our TodosView:

```
class TodosApp extends StatelessWidget {
  const TodosApp({
    Key? key,
    required this.todoController,
  }) : super(key: key);

  final TodosController todoController;

  @override
  Widget build(BuildContext context) {
    return MaterialApp(
      //...
      onGenerateRoute: (RouteSettings routeSettings) {
        return MaterialPageRoute<void>(
          settings: routeSettings,
          builder: (BuildContext context) {
            switch (routeSettings.name) {
              case TodosView.routeName:
                return TodosView(controller:
                                 todoController);
              default:
                return TodosView(controller:
                                 todoController);
            }
          },
        );
      },
    );
  }
}
```

In the preceding code, we are defining a TodoController as a required parameter to the TodosApp widget, and then forwarding the property to TodosView.

We haven't even begun to use the controller – we are just passing it down the widget tree. Once we inspect the TodosView widget, we can finally see how TodosController is being used:

```
class TodosView extends StatelessWidget {
  const TodosView({Key? key, required this.controller}) :
                   super(key: key);

  static const routeName = '/todos';
  final TodosController controller;
  ...
}
```

To start, observe that the controller is, once again, defined as a required parameter of TodosView. Now, let's take a closer look at the build function to see how TodosController is utilized:

```
class TodosView extends StatelessWidget {
  ...
  @override
  Widget build(BuildContext context) {
    return Scaffold(
      ...
      body: Container(
        ...
        child: ConstrainedBox(
          constraints: const BoxConstraints(maxWidth: 550),
          child: Column(
            children: [
              Padding(
                ...
                child: TodoTextField(
                  onSubmitted: (value) {
                    controller.createTodo(Todo(
                      id: controller.todos.length + 1,
                      task: value,
                    ));
                  },
                ),
```

```
              ),
              AnimatedBuilder(
                animation: controller,
                builder: (BuildContext context,
                        Widget? child) {
                  return ListView.builder(
                    shrinkWrap: true,
                    itemBuilder: (context, index) {
                      final todo = controller.todos[index];
                      return TodoListTile(
                        todo: todo,
                        onToggleComplete: (bool? value) {
                          controller.update(todo.copyWith(
                            complete: value));
                        },
                        onDelete: () {
                          controller.deleteTodo(todo);
                        },
                      );
                    },
                    itemCount: controller.todos.length,
                  );
                },
              ),
            ],
          ),
        ),
      ),
    );
  }
}
```

Notice that the controller is passed as a value to `AnimatedBuilder`. We also have several calls to functions that will create, update, and delete todos in our controller class.

Using AnimatedBuilder may seem strange because we are passing a ChangeNotifier as an animation, but upon inspecting the property, you will notice that the animation property is just a Listenable, which ChangeNotifier inherits from. Any time that the TodosController.notifyListeners method notifies its listeners of changes, the builder method of AnimatedBuilder will be re-rendered, ensuring that these changes only trigger efficient UI updates.

This process probably feels cumbersome because we had to add boilerplate code to get to the interesting parts of our code base – the business logic. Understanding the cumbersome nature of property passing will allow us to understand the benefits of InheritedWidget. Next, we'll refactor the same example to work with InheritedNotifier.

Creating an InheritedNotifier

To provide our TodoController without passing it down the tree manually, we will create a class that extends from InheritedWidget. Specifically, we will use a special type of InheritedWidget called InheritedNotifier that accepts a Listenable as its main parameter.

An InheritedNotifier is a special variant of InheritedWidget that accepts a Listenable subclass and updates its dependencies whenever the notifier is triggered.

To create a custom InheritedNotifier, create a class called TodoScope and add the following code:

```
class TodosScope extends InheritedNotifier<TodoController> {
  const TodosScope({
    Key? key,
    TodoController? notifier,
    required Widget child,
  }) : super(key: key, notifier: notifier, child: child);
  static TodoController of(BuildContext context) {
    return context.dependOnInheritedWidgetOfExactType
      <TodosScope>()!.notifier!;
  }
}
```

InheritedNotifier takes two parameters besides the key: a child widget and a notifier property. The latter inherits from Listenable. In the preceding code, we passed those three variables up to the parent constructor of TodosScope using the super keyword. Then, we defined a static method called TodosScope.of to retrieve our InheritedNotifier.

Inside this method, we use a function on `BuildContext` called `dependOnInheritedWidgetOfExactType`. This is a special function that obtains the nearest widget in the tree that matches the type we define, which must be the type of a concrete `InheritedWidget` subclass. Additionally, the function registers `BuildContext` so that changes to our `TodoScope` will trigger a rebuild to provide the new values to the widget.

Now that we have created our custom `InheritedWidget`, let's learn how to use our newly created `TodoScope`.

Getting information from TodosController using TodosScope

Next, let's refactor the logic in the `TodosView` widget so that it relies on `TodoScope` and remove the instances where `TodosController` is being passed as a prop. First, remove the `TodosController` property definition and delete it from the constructor:

```
class TodosView extends StatelessWidget {
  const TodosView({Key? key}) : super(key: key);

  static const routeName = '/todos';
  . . .
}
```

Next, change `AnimatedBuilder` to `Builder` and add the following code to access `TodosController` from `TodosScope`:

```
@override
Widget build(BuildContext context) {
  return Scaffold(
    . . .
    body: Container(
      alignment: Alignment.topCenter,
      child: ConstrainedBox(
        constraints: const BoxConstraints(maxWidth: 550),
        child: Column(
          children: [
            Builder(
              builder: (BuildContext context) {
                final controller =
                  TodosScope.of(context);
                  . . .
```

```
                        },
                    ),
                ],
            ),
          ),
        ),
      );
  }
```

Finally, wrap `ListView.builder` with a `Column` and move `TodoTextField` into `Column`. Additionally, add `shrinkWrap: true` to `ListView.builder` so that there are no rendering errors from nesting a `ListView` inside a `Column`:

```
          return Column(
            children: [
              Padding(
                padding:
                  const EdgeInsets.symmetric(
                    horizontal: 24,
                    vertical: 16,
                  ),
                child: TodoTextField(
                  onSubmitted: (value) {
                    controller.createTodo(Todo(
                      id:
                      controller.todos.length + 1,
                      task: value,
                    ));
                  },
                ),
              ),
              ...
            ],
          );
```

Let's look at the changes we've made in this refactor:

- We've moved the `TodosController` property from the `TodosView` constructor.
- We've refactored `AnimatedBuilder` to `Builder` to access `TodosController` from the `TodosScope` inherited widget.
- We've wrapped `ListView.builder` with a `Column` and added `TodoTextField` to `Column.children` to reuse `TodosController`.

If we run the following sample now, we will be met with a null check operator error, as shown here:

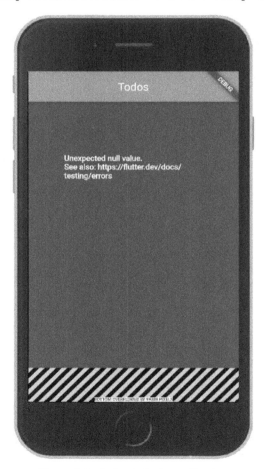

Figure 3.6 – A simple Todo application

The somewhat cryptic error displayed in *Figure 3.6* happened because Flutter failed to find an instance of `TodosScope` in the application's widget tree.

Since we have not provided this widget anywhere in the tree, the error we are seeing makes sense, but we could make the error a bit more descriptive. Before moving on to the next step, let's clean this up a bit. Change the `TodosController.of` method to the following:

```
static TodosController of(BuildContext context) {
  final TodosScope? result =
      context.
dependOnInheritedWidgetOfExactType<TodosScope>();
  assert(result != null,
         'No TodosScope found in context');
  assert(result!.notifier != null,
         'No TodosController found in context');
  return result!.notifier!;
}
```

Here, we have changed our helper method to display an error message when `TodosScope` is not found in the widget tree or when it does not include our notifier. Now, run the example and observe a message similar to what is shown here:

Figure 3.7 – A simple Todo application

Here, we can see the new, more descriptive error that's displayed when we run the application: **"No TodosScope found in context"**.

Next, let's provide our dependencies in the Todo application.

Providing TodosController in the widget tree

Finally, we should add our `TodoController` to our widget tree so that our `TodosScope` can correctly retrieve the controller inside `TodosView`. Inside the `main.dart` file, change the main function to the following:

```
void main() async {
  final todoController = TodosController();
  runApp(TodosScope(
    notifier: todoController,
    child: const TodosApp(),
  ));
}
```

Notice that we no longer pass the controller in as a property of `TodosApp`. Instead, we wrap the `TodosApp` widget in the `TodosScope` widget that we previously defined while passing in our instance of `TodosController`.

Now, upon running our application, we will no longer see the error messages from the previous steps and should be met with our Todo application running like it was previously!

Next, let's take a look at the popular `Provider` package and refactor our application to reduce some of the boilerplate introduced by `InheritedWidget`.

Refactoring with Provider

`InheritedWidget` and its associated classes work great, but hopefully, the previous section of this chapter has shown you that while it is a useful tool in our utility belt, it is a low-level widget that also requires some boilerplate.

What if, instead of using a `Listenable`, we wanted to use a `Stream` or a `Future` or even multiple `Listenable` classes? What if we wanted to make changes to specific properties to trigger rebuilds in different places throughout the application? The logic we would need to provide in our `TodosScope` (or any other `InheritedWidget`) would become much more complex and unwieldy. Even worse, this logic takes us away from the business functionality of our application.

Fortunately, there is an outstanding library that exists called `provider` that is promoted by the Flutter team. The `Provider` package includes several wrappers around `InheritedWidget` called Providers that handle each of the aforementioned use cases.

Installing Provider

First, let's install `provider` by running the following command in a terminal in the project's root directory:

```
flutter pub add provider
```

This command will add the latest version of the provider package from `https://pub.dev/`, Dart's package registry. Now, we can import the `Provider` package where needed.

Getting information from TodosController using Provider

As `Provider` is just a wrapper around `InheritedWidget`, refactoring is a breeze; our previously defined `TodosController` works without any additional effort.

Let's refactor the logic in the `TodosView` widget to rely on `Provider` rather than `TodosScope`. First, change `TodosView` so that it accesses `TodosController` using the following code:

```
class TodosView extends StatelessWidget {
  ...
  @override
  Widget build(BuildContext context) {
    final TodosController controller = context.read();
    ...
  }
}
```

Using `context.read()` allows us to access the controller without rebuilding the entire `TodosView` widget tree if `TodosController` changes. Finally, refactor `Builder.builder` to access the controller by using `context.select()` instead of `ProviderScope`:

```
Builder(
  builder: (BuildContext context) {
    final todos = context.select(
      (TodosController m) => m.todos);


  },
),
```

Here, we are using `context.select()` to subscribe to changes to `TodosController.todos`. With these two refactors, we are done updating `TodosView`. To recap, the two changes we have made use special `extension` methods that are defined on `BuildContext` by the `Provider` package:

- At the beginning of the `build` function, we use a special `context.read` to retrieve a non-rebuilding instance of `TodosController`.

- At the beginning of our `Builder` widget's builder function, we use a special `select` function only to retrieve the todos from `TodosController` and subscribe to their changes.

The use of these two methods demonstrates the power of the `Provider` package since it allows us to efficiently retrieve references to both `TodosController` and its values.

To further demonstrate Provider's benefits, when running this code, we are met with an extremely helpful and verbose error because `TodosController` cannot be found in the application's widget tree:

```
The following ProviderNotFoundException was thrown building
TodosView(dirty):
Error: Could not find the correct Provider above this TodosView
Widget
This happens because you used a BuildContext that does not
include the provider
of your choice. There are a few common scenarios:
You added a new provider in your main.dart and performed a hot-
reload.
To fix, perform a hot-restart.
The provider you are trying to read is in a different route.
Providers are "scoped". So if you insert of provider inside a
route, then
other routes will not be able to access that provider.
You used a BuildContext that is an ancestor of the provider you
are trying to read.
Make sure that TodosView is under your MultiProvider/Provider.
This usually happens when you are creating a provider and
trying to read it immediately.
```

This error message provides the developer with the information that they would need to debug all of the scenarios in which a provider might be incorrectly accessed without first being provided in the widget tree.

Providing TodosController in the widget tree

Finally, let's resolve the errors from the previous step by using `Provider` to provide the widget in the application's widget tree.

In the `main.dart` file, change the main function to the following:

```
void main() async {
  final todoController = TodosController();
  runApp(ChangeNotifierProvider.value(
    value: todoController,
    child: const TodosApp(),
  ));
}
```

As you may have guessed, `ChangeNotifierProvider` is a special widget that accepts a `ChangeNotifier` as a property and provides that class to descendant `BuildContext` references.

The `value` method allows you to pass in a previously defined value, but if for some reason you needed to dynamically create providers – for instance, if providers are dynamically changing based on a property or a route – there is a `create` property method that can be provided to the default constructor of `ChangeNotifierProvider`.

Let's see what that would look like:

```
void main() async {
  runApp(ChangeNotifierProvider(
    create: (_) => TodosController(),
    child: const TodosApp(),
  ));
}
```

The change we've made here will now dynamically create a new instance of `TodosController` whenever the widget tree changes. This feature would be useful if `TodosController` had a dynamic dependency.

This example only grazes the surface of the powerful abstractions of the `Provider` package, which includes a long list of features:

- Simplified allocation/disposal of resources
- Lazy loading
- A vastly reduced boilerplate over making a new class every time

- Developer tooling friendly – using `Provider`, the state of your application will be visible in Flutter's developer tools

- A common way to consume these inherited widgets (see `Provider.of`/`Consumer`/ `Selector`)

- Increased scalability for classes with a listening mechanism that grows exponentially in complexity (such as `ChangeNotifier`, which is *O(N)* for dispatching notifications).

We have successfully refactored our Todo application to use `Provider`! In doing so, we eliminated the boilerplate that we introduced by using `InheritedWidget`, while also using a library that handles many edge cases that come up when using `InheritedWidget`.

Now that we have wrapped up our application, let's review what we have learned.

Summary

In this chapter, we reviewed a simple Todo application that used property passing to share business logic. Then, we refactored the application to use `InheritedWidgets` to share dependencies. Finally, we used the popular `Provider` package to reduce the boilerplate that was introduced with inherited widgets.

By learning how to use inherited widgets and `Provider`, you now know how to store and share dependencies that can be retrieved further down the tree, eliminating some of the issues of property drilling.

In the next chapter, you will learn how to build platform-specific user interfaces using the Material and Cupertino libraries.

Further reading

To learn more about the topics that were covered in this chapter, take a look at the following resources:

- *State management*: `https://flutter.dev/docs/development/data-and-backend/state-mgmt/simple`

- *Flutter: Creating your own Inherited Widgets*: `https://blog.gskinner.com/archives/2021/04/flutter-creating-your-own-inherited-widgets.html`

- *Making sense of all those Flutter Providers*: `https://medium.com/flutter-community/making-sense-all-of-those-flutter-providers-e842e18f45dd`

- *What is InheritedWidget and How it works in a Flutter??*: `https://flutteragency.com/inheritedwidget-in-flutter/`

4

Building a Native Settings Application Using Material and Cupertino Widgets

One of the best features of Flutter is that the framework gives you control over every pixel. Flutter ships with its own custom widget catalog, allowing you to define your own design system while also shipping both **Material Design** and **Cupertino** widgets. This capability is a departure from many other frameworks that choose to rely on **original equipment manufacturer** (**OEM**) widgets. The major benefit is that your app can be truly unique, using the high-level widget libraries of Google or Cupertino or a widget library of your own mechanizations.

In this chapter, first, we will examine how to use features of the Flutter framework to build platform-specific UIs. Then, we will learn how to build our **Settings** application using Apple theming with Flutter's cupertino package. Wrapping up, we will learn how to build the same application using Material theming with Flutter's material package.

So, in this chapter, you will cover the following topics:

- Building a platform-specific UI

- Building an iOS-specific UI with CupertinoApp

- Building an Android-specific UI with MaterialApp

Technical requirements

Make sure to have your Flutter environment updated to the latest version in the stable channel. Clone our repository and use your favorite IDE to open the Flutter project we've built at chapter_4/step_0_start.

The project that we'll build upon in this chapter can be found on GitHub: `https://github.com/PacktPublishing/Cross-Platform-UIs-with-Flutter/tree/main/chapter_4/step_0_start`.

The complete source code can be found on GitHub as well: `https://github.com/PacktPublishing/Cross-Platform-UIs-with-Flutter/tree/main/chapter_4/step_4_end`.

Understanding that everything is a widget

Flutter relies heavily on component-driven development, in which UIs are built from the *bottom up* with basic components progressively being assembled into more complex components and, eventually, entire pages. In Flutter, these components are called **widgets**.

If you are new to Flutter, you may not be aware of its layered cake architecture. What this means is that Flutter is a series of independent libraries that each depend on the underlying layer.

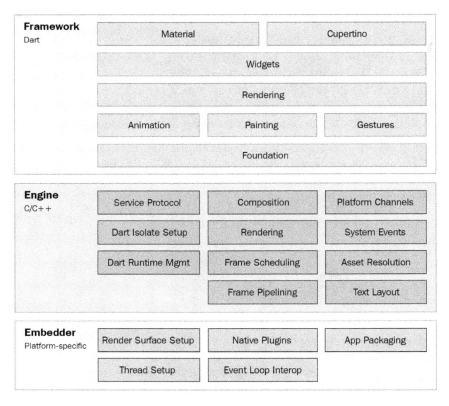

Figure 4.1 – Flutter's layered architecture (Source: https://docs.flutter.dev/resources/architectural-overview)

As you can see from *Figure 4.1*, the topmost libraries of the Framework layer are represented by Material and Cupertino widgets, which build upon a platform-agnostic set of widgets. An example of this in action would be `ElevatedButton` for Material and `CupertinoButton` for Cupertino, which compose widgets such as `Row`, `Align`, and `Padding` to implement their respective designs. The phrase *everything is a widget* denotes that every component used for structure, layout, navigation, or interaction is a widget.

One of the most exciting features of Flutter is that it ships the framework with the application, which enables the use of all the framework's widgets irrespective of the platform that the application is running on. Even common widgets defined under the `package:flutter/widgets.dart` library and `WidgetsApp` have platform-specific behaviors that can be overridden. `ListView`, for example, will customize its scrolling physics automatically; on an iOS device, reaching the scroll threshold results in a bouncing effect, whereas scrolling on an Android device results in a glow effect.

Let's explore how to set up the project so we can start to explore how to build a platform-specific UI in our application.

Setting up the project

In this chapter, we will learn how to use platform-specific widgets to build a simple Settings application. The application will accept user input for several groups of fields and allow the user to change their theme from the system setting to either light or dark mode.

When we have finished, the resulting application should mirror *Figure 4.2*, which shows the same application but with two different platform-specific designs:

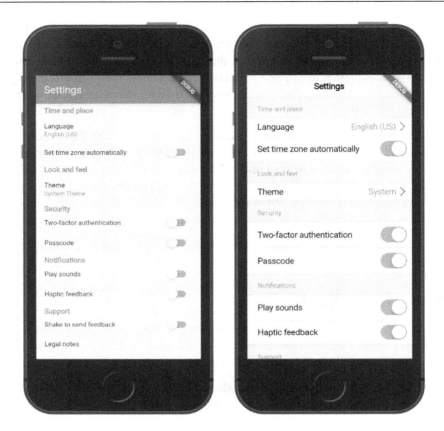

Figure 4.2 – Settings app

We will start with an example that displays an empty page and includes classes to update each application setting. We will first update this sample to display platform-specific messages to the user. We will then update the application to display a platform-specific UI when running on an Android or iOS device.

After downloading the initial application from GitHub, we can start by moving into its root directory and running the `flutter pub get` command from the terminal. This will install any package dependencies that the application needs. Additionally, opening the project in an IDE will also trigger an installation of the project on initial load and when the `pubspec.yaml` is updated.

After installing the application's dependencies, execute the `flutter run` command from the same terminal window and view the project either in your browser or on your currently running emulator. When viewing the application in your IDE, you should observe the following file structure:

```
lib
|-- src:
```

```
        |-- settings
        |-- app.dart
|-- main.dart
```

Let's briefly explore the purpose of each top-level folder/class:

- src: The source folder contains implementation details of our application that, unless exposed, will be hidden from consumers.

- settings: The settings folder contains SettingsController, which we will use to update each setting, and SettingsScope, which we will use to share the controller.

- app.dart: The widget that will control whether to display a Material-themed application or a Cupertino-themed application.

- main.dart: The entry point of the application.

Now, let's explore the features of the Flutter framework that enable us to build platform-specific UIs.

Building a platform-specific UI

In a Flutter application, there are two primary ways to access the platform that the application is running on: statically (using platform flags) or dynamically (using the TargetPlatform object from BuildContext). The platform can be detected statically using the list of platform flags provided by the Flutter framework, as detailed here:

- isLinux

- isMacOS

- isWindows

- isAndroid

- isIOS

- isFuschia

Each name implies the platform in which the application is executing. The tradeoff of using these flags is that they cannot be used to dynamically switch the styles for a platform. To do so, we turn to TargetPlatform defined on context. Before we dive into using MaterialApp to build our Settings application, let's demonstrate how to access the platform dynamically.

We can access the target platform information directly from BuildContext by using Theme. of(context).platform. Open the app.dart file and add the following code:

```
import 'package:flutter/material.dart';
```

```
class SettingsApp extends StatelessWidget {
  const SettingsApp({
    Key? key,
    this.platform,
  }) : super(key: key);

  final TargetPlatform? platform;

  @override
  Widget build(BuildContext context) {
    return MaterialApp(
      home: Scaffold(
        body: Column(
          mainAxisAlignment: MainAxisAlignment.center,
          children: [
            Text(platform?.name ?? 'Unknown'),
            if (platform == TargetPlatform.android)
              const Text('Hello Android'),
            if (platform == TargetPlatform.iOS)
              const Text('Hello iOS'),
          ],
        ),
      ),
    );
  }
}
```

There are three main refactors that occur in the previous code:

1. We create a platform property to be passed into the constructor.

2. We fall back to the platform from Theme.of(context).platform if one is not provided.

3. We render the text of the platform as well as a welcome message that is dependent on the platform.

Passing in the target platform allows us to dynamically change the platform from the application root. Let's test this by running the application on an Android emulator. We should see the following result:

Figure 4.3 – App targeting Android

The resulting code should render both the platform and a welcome message specific to Android matching *Figure 4.3*.

Now, change `main.dart`, passing `platform: TargetPlatform.iOS` into the constructor of `SettingsApp`, and run the same application. The results should match the following figure:

Figure 4.4 – App targeting iOS

Our application should now match *Figure 4.4*, displaying text that signifies the target platform is iOS rather than Android. Our ability to override even the platform information for the sake of displaying a different UI should drive home the capabilities of Flutter's widget system.

To fully enable `SettingsApp`, replace the code in the `build` function with the following:

```
@override
Widget build(BuildContext context) {
  final platform =
    this.platform ?? Theme.of(context).platform;
```

```
switch (platform) {
  case TargetPlatform.android:
  case TargetPlatform.fuchsia:
    return const MaterialSettingsApp();

  case TargetPlatform.iOS:
  case TargetPlatform.macOS:
  default:
    return const CupertinoSettingsApp();
  }
}
```

This code will return either `MaterialSettingsApp` or `CupertinoSettingsApp` depending on the target platform property. This code also demonstrates how we have the ability to display either UI regardless of platform – we could very well display `CupertinoSettingsApp` for Android.

Neither of these widgets has been created yet, but since we will start by defining `CupertinoSettingsApp`, comment out the line containing `MaterialSettingsApp` so that the iOS-specific Settings UI is always returned.

In the next section, you will see that regardless of whether the application is running on the Android emulator or the iOS simulator, the UI is still iOS-specific – a perfect demonstration of Flutter letting us choose the widgets to use regardless of the platform.

Now, let's learn how to build our iOS-specific Settings UI using Flutter's `cupertino` package.

Building an iOS-specific UI with CupertinoApp

Now that we have familiarized ourselves with Flutter's platform detection capabilities, let's put what we've learned into action by building a uniquely Cupertino Settings UI.

In Flutter, you can build an application that uses the iOS design system created by Apple by wrapping your application in `CupertinoApp`. This widget builds on `WidgetsApp`, which provides basic navigation and some foundational widgets by configuring the design system's standards.

Because `CupertinoApp` is just a widget, the design system that you decide to use is not restricted by platform. You could very well use Material for iOS applications and Cupertino for Android applications, or even build something completely custom!

As we want to use a generic theme mode to represent iOS, let's extend `SettingsService` first. Open `settings_service.dart` and replace the code with the following:

```dart
class SettingsService {
  Future<AdaptiveThemeMode> themeMode() async =>
    AdaptiveThemeMode.system;

  Future<void> updateThemeMode(AdaptiveThemeMode theme)
    async {
    //...
  }
}
```

```dart
enum AdaptiveThemeMode { system, light, dark }
```

We have introduced two significant changes in this refactor:

- At the bottom of the file, we defined an AdaptiveThemeMode.
- We replaced ThemeMode from the Material library with the one that we have defined.

Inside of the src directory, create a folder called cupertino. The first widget that we will create is CupertinoSettingsApp, which will configure CupertinoApp to use the theme mode defined in SettingsController. Then, create a file called cupertino_settings_app. dart and add the following code:

```dart
import 'package:flutter/cupertino.dart';
import 'package:settings/src/settings/settings_scope.dart';
import 'package:settings/src/settings/settings_service.dart';

import 'cupertino_settings_view.dart';

class CupertinoSettingsApp extends StatelessWidget {
  const CupertinoSettingsApp({Key? key}) : super(key: key);

  @override
  Widget build(BuildContext context) {
    final settingsController = SettingsScope.of(context);
    final brightness = settingsController.themeMode ==
      AdaptiveThemeMode.dark
        ? Brightness.dark
        : Brightness.light;
```

```
    return CupertinoApp(
      theme: CupertinoThemeData(
        brightness: brightness,
      ),
      home: const CupertinoSettingsView(),
    );
  }
}
```

Let's examine what this code does:

1. We created `CupertinoSettingsApp` as a stateless widget.
2. We accessed the theme from `SettingsController` in `CupertinoSettingsApp.build`.
3. We used the theme to set `CupertinoThemeData` brightness settings.

The brightness settings will determine whether our application uses a dark or light theme to render all the widgets that we will use from Flutter's `Cupertino` library.

Next, let's define the `CupertinoSettingsView` widget that `CupertinoSettingsApp` uses. Create a file called `cupertino_settings_view.dart`. In this file, we will define several private widgets to be used inside one main widget that is public.

To start, let's create `CupertinoSettingsView` with the following code:

```
class CupertinoSettingsView extends StatelessWidget {
  const CupertinoSettingsView({Key? key}) :
                              super(key: key);

  @override
  Widget build(BuildContext context) {
    return const CupertinoPageScaffold(
      navigationBar: CupertinoNavigationBar(
        middle: Text('Settings'),
      ),
      child: SafeArea(
        child: _SettingsForm(),
      ),
    );
```

```
    }
  }
```

In this code block, we are creating a stateless widget that builds a page scaffold using CupertinoPageScaffold and CupertinoNavigationBar. As a child of CupertinoPageScaffold, we pass in a _SettingsForm widget (which we will create next) wrapped in a SafeArea widget.

Let's add that _SettingsForm widget by introducing the following code:

```
class _SettingsForm extends StatelessWidget {
  const _SettingsForm({
    Key? key,
  }) : super(key: key);

  @override
  Widget build(BuildContext context) {
    final settingsController = SettingsScope.of(context);
    final theme = CupertinoTheme.of(context);

    return ListView(
      children: [],
    );
  }
}
```

In this code block, we are only retrieving SettingsController and the CupertinoTheme objects from context, then building ListView. We will use the classes that we retrieve here as we build out the rest of the page.

Running the code should produce the following result:

Figure 4.5 – Blank application

Our application should match *Figure 4.5*, displaying only an application bar.

Now, let's use CupertinoFormSection to add a section for time settings. Add the following code to the children of ListView:

```
CupertinoFormSection(
  header: const Text('Time and place'),
  children: [
    CupertinoFormRow(
      child: Row(
        mainAxisSize: MainAxisSize.min,
        children: [
```

```
            Text (
              "English (US)",
              style:
                theme.textTheme.textStyle.copyWith(
                color: theme.textTheme.textStyle
                  .color!.withOpacity(.5),
              ),
            ),
            const Icon(CupertinoIcons.chevron_right),
          ],
        ),
        prefix: const Text("Language"),
      ),
      CupertinoFormRow(
        child: CupertinoSwitch(
          value: true,
          onChanged: (value) {},
        ),
        prefix:
          const Text("Set time zone automatically"),
      ),
    ],
  ),
```

CupertinoFormSection can be used to create an iOS-style form section. Also, notice that we are using CupertinoFormRow to create two iOS-style form rows in this section:

- One is the **Language** form row, styled as an action list item that could navigate to another page to select languages.
- The other is the **Time Zone** form row, using CupertinoSwitch to adjust the time zone automatically.

Now, if we run the code with these updates, we should see the following result:

Figure 4.6 – Settings app with a Time and place section

Our application should match *Figure 4.6*, displaying a **Time and place** section with **Language** and time zone information.

Now, let's build a section that uses our `SettingsController`. While still in `ListView.children`, add the following code after the **Time and place** form section:

```
CupertinoFormSection(
        header: const Text("Look and feel"),
        children: [
          GestureDetector(
            onTap: () async {
              final themeMode =
                  await showCupertinoModalPopup
                    <AdaptiveThemeMode?>(
                context: context,
                builder: (context) {
```

```
                  return const _ThemeActionSheet();
                },
              );


              if (themeMode != null) {
                settingsController.updateThemeMode(
                  themeMode);
              }
            },
          },
          child: CupertinoFormRow(
            child: Row(
              mainAxisSize: MainAxisSize.min,
              children: [
                Text(
                  settingsController.themeMode.
                    name.capitalize,
                  style:
                    theme.textTheme.textStyle.copyWith(
                    color: theme.textTheme.textStyle
                      .color!.withOpacity(.5),
                    ),
                ),
                const Icon(
                  CupertinoIcons.chevron_right),
              ],
            ),
            prefix: const Text("Theme"),
          ),
        ),
      ],
    ),
```

Let's walk through each highlighted section and explain what's happening:

1. When the theme section is tapped, an action sheet that returns a theme mode will be displayed using showCupertinoModalPopup.

2. The `builder` function for `showCupertinoModalPopup` will return a `_ThemeActionSheet` widget, which we will define next.

3. The theme mode will be used to update the `SettingsController` theme mode.

4. The currently selected theme mode will be displayed in `CupertinoFormRow`.

To finish up this section, let's create our `_ThemeActionSheet` widget by adding the following code at the bottom of the file:

```
class _ThemeActionSheet extends StatelessWidget {
  const _ThemeActionSheet({
    Key? key,
  }) : super(key: key);

  @override
  Widget build(BuildContext context) {
    return CupertinoActionSheet(
      title: const Text('Choose app theme'),
      actions: [
        CupertinoActionSheetAction(
          child: const Text('System Theme'),
          onPressed: () {
            Navigator.pop(context,
                          AdaptiveThemeMode.system);
          },
        ),
        CupertinoActionSheetAction(
          child: const Text('Light Theme'),
          onPressed: () {
            Navigator.pop(context,
                          AdaptiveThemeMode.light);
          },
        ),
        CupertinoActionSheetAction(
          child: const Text('Dark Theme'),
          onPressed: () {
            Navigator.pop(context, AdaptiveThemeMode.dark);
          },
```

```
          ),
        ],
        cancelButton: CupertinoActionSheetAction(
          child: const Text('Cancel'),
          onPressed: () {
            Navigator.pop(context);
          },
        ),
      );
    }
  }
```

In this example, our widget builds a list of theme options using the `CupertinoActionSheet` and `CupertinoActionSheetAction` widgets. If any `CupertinoActionSheetAction` is pressed, we use `Navigator.pop` to close the modal and return the selected theme.

Before running the code, return to `SettingsApp` and uncomment the line that returns `MaterialSettingsApp`. Now, when we run this code on the Android emulator, we should see the following result:

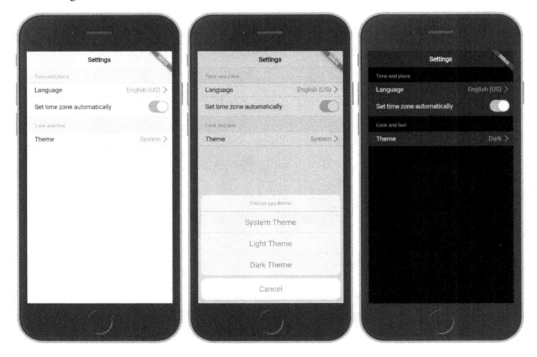

Figure 4.7 – Settings application with theme section

Our application should match *Figure 4.7*, which demonstrates our newly added **Look and feel** section, the action sheet that is displayed, and the result of selecting **Dark Theme** in the theme selector.

To finish up, let's add form sections for **Security**, **Notifications**, and **Support** so that our Settings application will be complete. To do this, add the following code to `ListView.children` for the `Security` settings:

```
CupertinoFormSection(
  header: const Text("Security"),
  children: [
    CupertinoFormRow(
      child: CupertinoSwitch(
        value: true,
        onChanged: (value) {},
      ),
      prefix:
        const Text("Two-factor authentication"),
    ),
    CupertinoFormRow(
      child: CupertinoSwitch(
        value: true,
        onChanged: (value) {},
      ),
      prefix: const Text("Passcode"),
    ),
  ],
),
```

Follow up that block with another block for `Notifications`:

```
CupertinoFormSection(
  header: const Text("Notifications"),
  children: [
    CupertinoFormRow(
      child: CupertinoSwitch(
        value: true,
        onChanged: (value) {},
      ),
      prefix: const Text("Play sounds"),
```

```
          ),
        CupertinoFormRow(
          child: CupertinoSwitch(
            value: true,
            onChanged: (value) {},
          ),
          prefix: const Text("Haptic feedback"),
        ),
      ],
    ),
```

And finally, add the last block for Support:

```
          CupertinoFormSection(
            header: const Text("Support"),
            children: [
              CupertinoFormRow(
                child: CupertinoSwitch(
                  value: true,
                  onChanged: (value) {},
                ),
                prefix: const Text("Shake to send feedback"),
              ),
              CupertinoFormRow(
                child: Row(
                  mainAxisSize: MainAxisSize.min,
                  children: const [
                    Icon(CupertinoIcons.chevron_right),
                  ],
                ),
                prefix: const Text("Legal notes"),
              ),
            ],
          ),
```

This code uses the same techniques and widgets that we just learned to build out the rest of the **Settings** UI.

Now, if we run the application, we should be met with the following results:

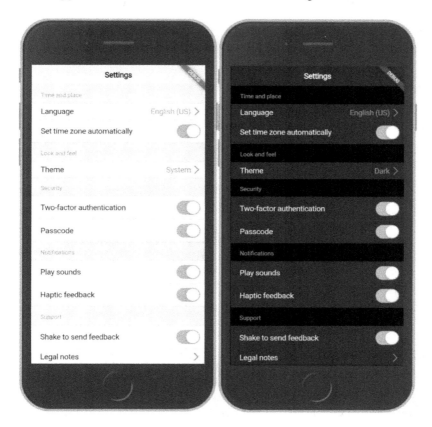

Figure 4.8 – Settings app in light/dark mode

If the application matches *Figure 4.8*, we should see a full set of settings and be able to toggle the application theme from being automatically detected by the system platform to manually set to light/dark.

In the next section, we will pivot to building our Android-specific Settings UI using `MaterialApp` widgets.

Building an Android-specific UI with MaterialApp

Now that we have built the iOS-specific version of our Settings app, we can focus on the Android-specific UI. Like `CupertinoApp`, `MaterialApp` builds on top of `WidgetsApp` by configuring the iOS design system created by Apple.

Inside of the `src` directory, create a folder called `material`. The first widget that we will create is our `MaterialSettingsApp`, which will configure `MaterialApp` to use the theme mode defined in `SettingsController`. Create a file called `material_settings_app.dart` and add the following code:

```dart
class MaterialSettingsApp extends StatelessWidget {
  const MaterialSettingsApp({Key? key}) : super(key: key);

  @override
  Widget build(BuildContext context) {
    final settingsController = SettingsScope.of(context);

    return MaterialApp(
      theme: ThemeData(),
      darkTheme: ThemeData.dark(),
      themeMode:
        settingsController.themeMode.materialThemeMode,
      home: const MaterialSettingsView(),
    );
  }
}

extension ThemeModeExtension on AdaptiveThemeMode {
  ThemeMode get materialThemeMode {
    switch (this) {
      case AdaptiveThemeMode.dark:
        return ThemeMode.dark;
      case AdaptiveThemeMode.light:
        return ThemeMode.light;
      default:
        return ThemeMode.system;
    }
  }
}
```

This code mirrors the behavior of the CupertinoSettingsApp widget:

1. We created a MaterialSettingsApp as a stateless widget.

2. We accessed the theme from SettingsController in MaterialSettingsApp. build.

3. We used themeMode to set the ThemeData brightness settings, adding an extension to convert AdaptiveThemeMode to ThemeMode.

The themeMode setting will determine whether our application uses a dark or light theme to render all the widgets that we will use from Flutter's material package. This is the first major difference between the material and cupertino packages that relies on the brightness setting.

Next, let's define the MaterialSettingsView widget that MaterialSettingsApp uses. Create a file called material_settings_view.dart in the same folder. In this file, we will define several private widgets to be used inside one main widget that is public.

To start, let's create MaterialSettingsView with the following code:

```
class MaterialSettingsView extends StatelessWidget {
  const MaterialSettingsView({Key? key}) : super(key: key);

  @override
  Widget build(BuildContext context) {
    return Scaffold(
      appBar: AppBar(
        title: const Text('Settings'),
      ),
      body: _SettingsForm(),
    );
  }
}
```

In this code block, we are creating a stateless widget that builds a page scaffold using Scaffold and AppBar. As a child of Scaffold, we passed in a _SettingsForm widget, which we will create next.

Let's add that _SettingsForm widget by introducing the following code:

```
class _SettingsForm extends StatelessWidget {
  const _SettingsForm({
    Key? key,
```

```
  }) : super(key: key);

  @override
  Widget build(BuildContext context) {
    final controller = SettingsScope.of(context);

    return ListView(
      children: [],
    );
  }
}
```

If you are experiencing déjà vu, that is probably because this code is identical to `_SettingsForm` used in `CupertinoSettingsView`. The repetition is fine – we are not trying to prematurely abstract or optimize our code before we understand the patterns.

Before we start building out our form, we will define a helper widget called `MaterialFormSection` that approximates some of the capabilities of `CupertinoFormSection`.

In the same file, add the following code at the bottom:

```
class MaterialFormSection extends StatelessWidget {
  const MaterialFormSection({
    Key? key,
    this.header,
    required this.children,
  }) : super(key: key);

  final Widget? header;
  final List<Widget> children;

  @override
  Widget build(BuildContext context) {
    assert(children.isNotEmpty);

    return Padding(
      padding: const EdgeInsets.only(left: 16, right: 16,
                                     top: 6),
      child: Column(
```

```
          crossAxisAlignment: CrossAxisAlignment.start,
          children: [
            if (header != null)
              Align(
                alignment: AlignmentDirectional.centerStart,
                child: DefaultTextStyle(
                  style: TextStyle(
                    color:
                      Theme.of(context).colorScheme.secondary,
                    fontWeight: FontWeight.bold,
                  ),
                  child: Container(
                    child: header,
                  ),
                ),
              ),
            ...children,
          ],
        ),
      );
    }
}
```

Here, we have done the following:

1. Defined a MaterialFormSection widget that accepts a header widget and children widgets as properties.

2. Asserted that the children list is not empty.

3. Added a padded Column to position the elements horizontally.

4. Added a center-left aligned header if the header property is present with special theming.

5. Added the children list to the end of Column.

Now we are ready to use the MaterialFormSection widget that we have created. Add the following code to the children of ListView inside the _SettingsForm widget:

```
MaterialFormSection(
  header: const Text('Time and place'),
  children: [
```

```
            const ListTile(
              contentPadding: EdgeInsets.zero,
              title: const Text('Language'),
              subtitle: Text("English (US)"),
              dense: true,
            ),
            SwitchListTile(
              title:
                const Text('Set time zone automatically'),
              contentPadding: EdgeInsets.zero,
              dense: true,
              value: controller.enableAutoTimeZone,
              onChanged:
                controller.updateEnableAutoTimeZone,
            ),
          ],
        ),
```

In this code, we have done the following:

1. Added `MaterialFormSecton` for the **Time and place** section to create an Android-style form section.

2. Added `ListTile` to create an Android-style form row for language in this section.

3. Added `SwitchListTile` to create an Android-style switch for this section.

Now, if we run the code with these updates, we should see the following result:

Figure 4.9 – Material Settings app in light/dark mode

Our application should match *Figure 4.9*, displaying adding a new section with the form rows for **Language** and time zone.

Now, let's build a section that uses `SettingsController`. While still in `ListView.children`, add the following code after the **Time and place** form section:

```
MaterialFormSection(
        header: const Text('Look and feel'),
        children: [
          ListTile(
            contentPadding: EdgeInsets.zero,
            title: const Text('Theme'),
            subtitle: Text(_themeDisplayTextMap[
              controller.themeMode] ?? ''),
            dense: true,
            onTap: () async {
```

```
              var themeMode = await
                showModalBottomSheet<AdaptiveThemeMode?>(
                  context: context,
                  builder: (context) {
                    return const _ThemeBottomSheet();
                  });

              if (themeMode != null) {
                controller.updateThemeMode(themeMode);
              }
            },
          ),
        ],
      ),
```

Let's walk through each highlighted section and explain what's happening:

1. When the theme section is tapped, an action sheet that returns a theme mode will be displayed using showModalBottomSheet.

2. The builder function for showModalBottomSheet will return a _ThemeBottomSheet widget, which we will define next.

3. The theme mode will be used to update the SettingsController theme mode.

4. The currently selected theme mode will be displayed in ListTile.

To finish up this section, let's create our _ThemeBottomSheet widget by adding the following code at the bottom of the file:

```
class _ThemeBottomSheet extends StatelessWidget {
  const _ThemeBottomSheet({
    Key? key,
  }) : super(key: key);

  @override
  Widget build(BuildContext context) {
    return Wrap(
      children: [
        ListTile(
          title: const Text('System Theme'),
```

```
              dense: true,
              onTap: () {
                Navigator.pop(context,
                              AdaptiveThemeMode.system);
              },
            ),
            ListTile(
              title: const Text('Light Theme'),
              dense: true,
              onTap: () {
                Navigator.pop(context,
                              AdaptiveThemeMode.light);
              },
            ),
            ListTile(
              title: const Text('Dark Theme'),
              dense: true,
              onTap: () {
                Navigator.pop(context, AdaptiveThemeMode.dark);
              },
            )
          ],
        );
      }
}
```

In this example, our widget builds a list of theme options using the `Wrap` and `ListTile` widgets. If any `ListTile` is pressed, we use `Navigator.pop` to close the modal and return the selected theme.

When we run this code, we should see the following result:

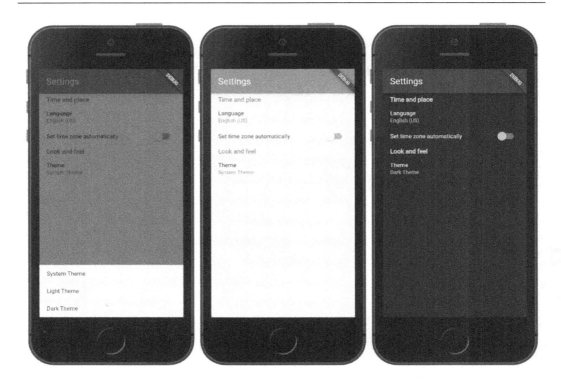

Figure 4.10 – Material Settings application with theme section

Our application should match *Figure 4.10*, which demonstrates our newly added **Theme** section (the bottom sheet that is displayed) and the result of selecting **Dark Theme** in the theme selector.

To finish up, let's add form sections for **Security**, **Notifications**, and **Support** so that our Settings application will be complete. Add the following code to `ListView.children`:

```
MaterialFormSection(
  header: const Text('Security'),
  children: [
    SwitchListTile(
      title:
        const Text('Two-factor authentication'),
      contentPadding: EdgeInsets.zero,
      dense: true,
      value:
        controller.enableTwoFactorAuthentication,
      onChanged: controller
```

```
                .updateEnableTwoFactorAuthentication,
        ),
        SwitchListTile(
          title: const Text('Passcode'),
          contentPadding: EdgeInsets.zero,
          dense: true,
          value: controller.enablePasscode,
          onChanged: controller.updateEnablePasscode,
        ),
      ],
    ),
    MaterialFormSection(
      header: const Text('Notifications'),
      children: [
        SwitchListTile(
          title: const Text('Play sounds'),
          contentPadding: EdgeInsets.zero,
          dense: true,
          value: controller.enableSounds,
          onChanged: controller.updateEnableSounds,
        ),
        SwitchListTile(
          title: const Text('Haptic feedback'),
          contentPadding: EdgeInsets.zero,
          dense: true,
          value: controller.enableHapticFeedback,
          onChanged:
              controller.updateEnableHapticFeedback,
        ),
      ],
    ),
    MaterialFormSection(
      header: const Text('Support'),
      children: [
        SwitchListTile(
          title: const Text('Shake to send feedback'),
```

```
          contentPadding: EdgeInsets.zero,
          dense: true,
          value: controller.enableSendFeedback,
          onChanged:
            controller.updateEnableSendFeedback,
        ),
        ListTile(
          title: const Text('Legal notes'),
          contentPadding: EdgeInsets.zero,
          dense: true,
          onTap: () {},
        ),
      ],
    ),
```

This code uses the same techniques and widgets that we just learned to build out the rest of the Settings UI.

Now, if we run the application, we should be met with the following results:

Figure 4.11 – Settings app in light/dark mode

If the Material application matches *Figure 4.11*, we should see a full set of settings and be able to toggle the application theme from being automatically detected by the system to manually set to **Light/Dark**.

Now that we have wrapped up our application, let's review what we have learned.

Summary

In this chapter, we learned how to implement a platform-specific UI by using `TargetPlatform` defined on context. Then, we created an iOS-specific UI by using widgets from Flutter's `cupertino` package. Finally, we recreated the same application with an Android-specific UI by using widgets from Flutter's `material` package.

By learning how to build platform-specific applications, you now understand how to build applications that adapt to the platforms that they run on.

In the next chapter, you will learn how to bring your application to life using animations.

Further reading

- *Platform-specific behaviors and adaptations*: `https://docs.flutter.dev/resources/platform-adaptations`

- *Building a Cupertino app with Flutter*: `https://codelabs.developers.google.com/codelabs/flutter-cupertino`

- *Detecting the Current Platform – iOS or Android*: `https://flutterigniter.com/detect-platform-ios-android`

5

Exploring Navigation and Routing with a Hacker News Clone

So far in our adventures, we have yet to build more than a simple application, however in real life, most of the applications that we use in our daily lives have more than one page. Due to this, every framework has a way of navigating between different views or pages; this feature allows us to keep the user's experience simple and intuitive because we only display the information that they need to see rather than everything that might lead to information overload.

In this chapter, we will learn how to navigate in a multi-page application using Flutter and its `Navigator` APIs. First, we will examine how navigation works in Flutter and how we can move from one screen of an application to another. Then, we will build a news application with the Navigator 1.0 API. After getting a clear understanding of how Navigator 1.0 works, we will refactor our application to use Navigator 2.0. Finally, we will simplify our application by using a popular routing library.

We will cover the following topics:

- Understanding navigation
- Imperative routing with Navigator 1.0
- Declarative routing with Navigator 2.0
- Simplifying Navigator 2.0 with GoRouter

Technical requirements

Make sure to have your Flutter environment updated to the latest version in the `stable` channel. Clone our repository and use your favorite IDE to open the Flutter project we've built at `chapter_5/step_0_start`.

The project that we'll build upon in this chapter can be found on GitHub: `https://github.com/PacktPublishing/Flutter-UI-Projects-for-iOS-Android-and-Web/tree/main/chapter_5/step_0_start`.

The complete source code can be found on GitHub as well: `https://github.com/PacktPublishing/Cross-Platform-UIs-with-Flutter/tree/main/chapter_5/step_4_end`.

Let's set up the project so we can start to explore how navigation works in a multi-page Flutter application.

Setting up the project

In this chapter, we will learn how to use Flutter's navigation APIs to build an interactive, multi-page Hacker News application that will allow us to navigate from lists of articles or stories, to their detail pages. Hacker News is a social news website focusing on computer science and entrepreneurship.

When we have finished, the resulting application should mirror *Figure 5.1*:

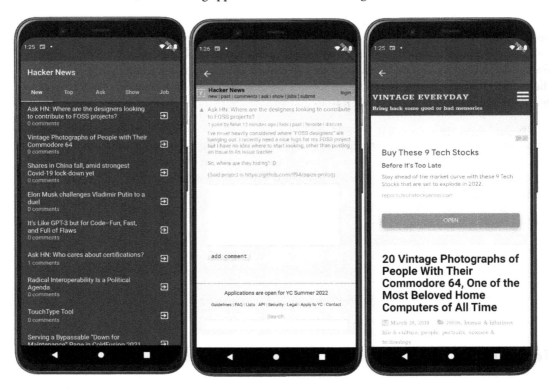

Figure 5.1 – Hacker News app

In *Figure 5.1*, we see three different screens:

- A home screen that has tabs for each *Story Type*
- A details or comments screen that renders Hacker News in a `WebView`
- A story link view that renders the original story source URL in a `WebView`

Additionally, when tapping on an already selected tab from either the details or the story screen, we should be redirected back to the list view for that *Story Type*.

We will start with an example that has one empty page with tabs that match the Hacker News web application at `https://news.ycombinator.com`: **New**, **Past**, **Comments**, **Ask**, **Show**, and **Jobs**. This example also includes the controllers and APIs to communicate with the Hacker News API located at `https://hacker-news.firebaseio.com`.

After downloading the initial application from GitHub, we can start by moving into its root directory and running the `flutter pub get` command from the Terminal. This will install any package dependencies that the application needs. Additionally, opening the project in an IDE will also trigger an install of the project on initial load and when `pubspec.yaml` is updated.

After installing the application's dependencies, execute the `flutter run` command from the same terminal window and view the project either in your browser or on your currently running emulator. When viewing the application in your IDE, you should observe the following file structure:

```
lib
|-- src:
   |-- data
   |-- localization
   |-- ui
      |-- story
   |-- app.dart
|-- main.dart
```

Let's briefly explore the purpose of each top-level folder/class:

- `src`: This folder contains implementation details of our application that, unless exposed, will be hidden from consumers.
- `data`: This folder that includes all of the business components of our application. This can include API facades.
- `localization`: This folder that includes all of the translations for the application.
- `ui`: This folder that includes all of the presentational components of our application. This can include controllers and widgets.

- `story`: This folder that includes all of the presentational components associated with stories.
- `app.dart`: This widget will control whether to display a Material-themed application or a Cupertino-themed application.
- `main.dart`: This is the entry point of the application.

Now let's explore the features of the Flutter framework that enable us to build platform-specific UIs.

Understanding navigation

Flutter has two ways of navigating between views. The first pattern introduced was an imperative API that makes use of the `Navigator` widget. Think of `Navigator` as a stack that sits at the top of the application. A stack is a linear data structure that follows the principle of **Last In First Out (LIFO)**, meaning the last element inserted inside the stack is removed first. Stacks generally have two main functions: push which allows you to add a new item to the stack, and `pop`, which removes the item from the stack that was added. The `Navigator` widget extends this API further:

- `pop`: Pops the top-most route off the navigator.
- `popAndPushNamed`: Pops the top-most route off the navigator and pushes a named route onto the navigator.
- `popUntil`: Pops routes off the navigator until some conditions are met.
- `Push`: Pushes a route onto the navigator.
- `pushAndRemoveUntil`: Pushes a route onto the navigator and pops routes until some conditions are met.
- `pushNamed`: Pushes a named route onto the navigator.
- `pushNamedAndRemoveUntil`: Pushes a named route onto the navigator and pops routes until some conditions are met.
- `pushReplacement`: Replaces the current route by first pushing the given route and then disposing of the previous one once the animation has finished.
- `pushNamedReplacement`: Replaces the current route by first pushing the given named route and then disposing of the previous one once the animation has finished.

In this chapter, we will primarily be focused on `pop` and `pushNamed`. Like any other widget in Flutter, `Navigator` can be deeply nested inside of other navigators, and when you make an API call, it will only affect the `Navigator` widget most closely associated with the given context.

To tell Flutter how to define the onGenerateRoute property on MaterialApp. If you inspect app.dart, you will notice that we have already done this:

```
onGenerateRoute: (RouteSettings routeSettings) {
  return MaterialPageRoute<void>(
    settings: routeSettings,
    builder: (BuildContext context) {
      switch (routeSettings.name) {
        case StoryDetailsView.routeName:
          return const StoryDetailsView(
            storyId: 0,
          );
        case StoryListView.routeName:
        default:
          return const StoryListView();
      }
    },
  );
},
```

The previous code snippet does the following:

- Define a MaterialPageRoute with the routeSettings provided by onGenerateRoute.
- Display the StoryDetailsView widget if routeSettings.name matches StoryDetailsView.routeName.
- Display the StoryListView widget if routeSettings.name matches StoryListView.routeName.

If we run our application, we should get an image that matches the following:

Figure 5.2 – Story list view

Notice in *Figure 5.2* that we have no way of getting to the StoryDetailsView. Let's add that ability by opening StoryListView and adding the following block of code to the Column's children:

```
TextButton(
    onPressed: () {
        Navigator.of(context).pushNamed(
            StoryDetailsView.routeName);
```

```
    },
    child: const Text('Go to details'),
)
```

Let's examine what this code does:

- We created a button with the text **Go to details**.

- On button press, we call `Navigator.pushNamed` and pass it a static string defined on `StoryDetailsView`.

Now if we reload the application, we should get our newly created button below the **Story List** text:

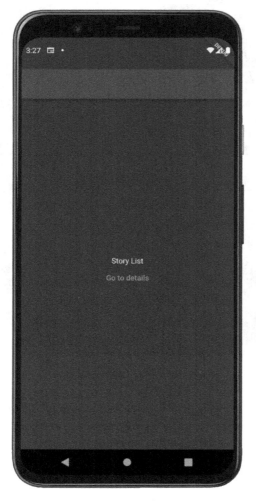

Figure 5.3 – Story list view with navigation button

If we click on our newly added button, as displayed in *Figure 5.3*, we should be redirected to the **Story Details** page and see the following screen:

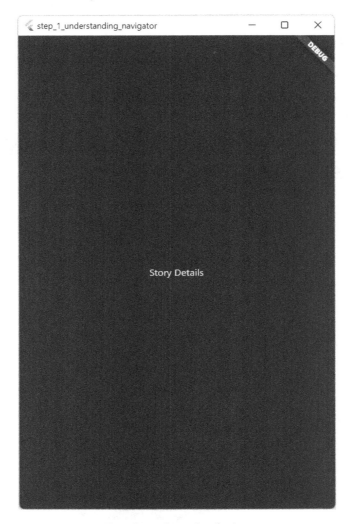

Figure 5.4 – Story details view

One important thing to notice in *Figure 5.4* is that we have no way of navigating back to the page that we came from. Let's add a button to return to our previous page by extending `StoryDetailsView` with the following block of code to the Column's children:

```
TextButton(
    onPressed: () {
```

```
        Navigator.of(context).pop();
    },
    child: const Text('Go back'),
),
```

Let's examine what this code does:

- We created a button with the text **Go back**.

- On button press, we call `Navigator.pop` which will remove the most recent route from the stack.

Upon refreshing the application, we should see our newly added **Go back** button. Our **Story Details** view should mirror the following screen:

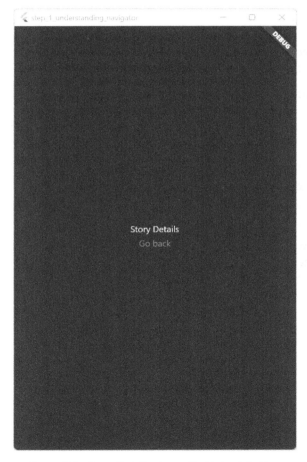

Figure 5.5 – Story Details view with Go back button

Graphically speaking, the example in *Figure 5.5* is fine, but technically, always displaying a back button on a screen can create a bad user experience. If somehow the user navigates to this screen initially, there would be no previous route to navigate to, and clicking on the button could cause an error.

Fortunately, the `Navigator.canPop` function allows us to verify that the stack has a previous page to navigate to. Let's wrap our button with a conditional check using the `Navigator.canPop` function using the following code:

```
if (Navigator.canPop(context))
    TextButton(
    onPressed: () {
        Navigator.of(context).pop();
    },
    child: const Text('Go back'),
),
```

In this refactored code snippet, we instruct the application to only display the back button if `Navigator.canPop` is `true`, signifying that there are additional routes on the stack beyond the current route.

You may be thinking that handling back navigation seems like an unnecessary step, and you would be correct. Flutter already has the ability to automatically imply whether or not to display a back button. This capability is built into `AppBar`, which will determine whether or not to display a back button if the leading property is empty and the `automaticallyImplyLeading` property is `true`.

Let's add `AppBar` to both `StoryListView` and `StoryDetailsView` to see what this looks like. Extend the `Scaffold` widget in both classes with the following code before `body`:

```
appBar: AppBar(),
```

The preceding code merely tells the application that we would like to use Flutter's Material `AppBar` with all of its default settings. These defaults include a `leading` property being empty and `automaticallyImplyLeading` being set to `true`. If we refresh our application, we should get the following results:

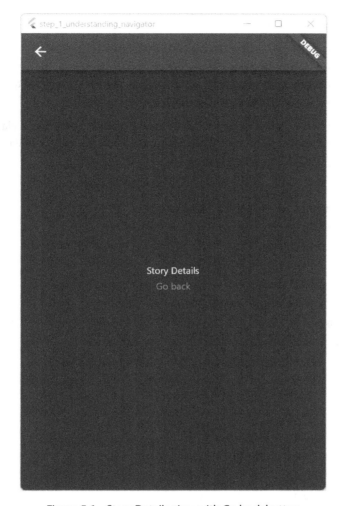

Figure 5.6 – Story Details view with Go back button

Notice in *Figure 5.6* that the top area of the application now includes a short bar that is empty on the `StoryListView`, and has a back button signified by a left arrow icon on `StoryDetailView`. The `AppBar` widget automatically injects that widget based on the navigator stack.

Now we should have a clearer understanding of the setup that is required when using Flutter's `Navigator`. We should also be a bit more comfortable using some of the imperative `Navigator`'s APIs to navigate between pages of an application. Next, let's build our Hacker News application using Flutter's `Navigator 1.0` APIs.

Imperative routing with Navigator 1.0

Now that we have familiarized ourselves with how to do multi-page navigation in Flutter, let's put what we've learned into action by building our Hacker News Application using these APIs. Flutter has a collection of APIs for handling multi-page navigation. The legacy APIs are referred to as `Navigator 1.0`. Let's start by using those to build our page navigation.

We will start by adding a new view with tabs. Create a new `home_view.dart` class in the `lib/src/ui/home` folder. In it create a new `StatefulWidget` called `HomeView` and add the following logic to the `_HomeViewState` class' `build` method:

```
@override
Widget build(BuildContext context) {
  return Scaffold(
    appBar: AppBar(
      title: const Text('Hacker News'),
      bottom: TabBar(
        controller: _controller,
        tabs: [
          for (final storiesType in StoriesType.values)
            Tab(text: storiesType.displayName)
        ],
      ),
    ),
    body: TabBarView(
      controller: _controller,
      children: [
        for (final storiesType in StoriesType.values)
          StoryListView(
            storiesType: storiesType,
          ),
      ],
    ),
  );
}
```

In the previous code snippet, we have done the following:

- Constructed a page that displays an AppBar which will display tabs from the StoriesType enum values.
- Constructed Scaffold.body with TabBarView that creates StoryListView for each StoriesType enum.
- Additionally, we pass both TabBar and TabBarView a TabController named _controller that we have yet to define.

Let's do so now by adding the following code in the same _HomeViewState class before the build method:

```
late final TabController _controller;

@override
void initState() {
  super.initState();
  _controller = TabController(
    length: StoriesType.values.length,
    vsync: this,
  );
}

@override
void dispose() {
  _controller.dispose();
  super.dispose();
}
```

Now, we have done the following:

- Instantiated an instance of our _controller.
- Created our _controller appropriately in our initState function.
- Disposed of the _controller in the dispose function.

Re-run the application and observe changes, which should match the following screen:

Figure 5.7 – Home view with tabs

Figure 5.7 shows us our newly created tabs with the pre-existing `StoryListView`. Now let's populate `StoryListView` with some actual data. Navigate to the widget and convert it from a `StatelessWidget` to a `StatefulWidget`. Replace the `Selector.builder` function with the following code:

```
return ListView.builder(
   itemCount: state.length,
   itemBuilder: (BuildContext context,
                int index) {
     final story = stories[index];
```

```
            return ListTile(
              title: Text(
                story.title,
              ),
              subtitle: GestureDetector(
                onTap: () =>
                  _navigateToStoryComments(story.id),
                child: Text('${story.descendants}
                          comments'),
              ),
              trailing: IconButton(
                onPressed: () =>
                  _navigateToWebview(story.id),
                icon: const Icon(Icons.exit_to_app),
              ),
            );
          },
        );
```

The previous code will use the stories provided by the `Selector` widget and build a list from those results. Additionally, we need to add the call to `getStories` to load the stories from the API.

Next, modify the `initState` method and replace it with the following code:

```
@override
void initState() {
  super.initState();
  _storyController = StoryController(context.read())
    ..getStories(widget.storiesType);
}
```

In this code, we immediately call the `getStories` method after initializing our controller. Finally, to run the application let's add some empty methods for `_navigateToStoryComments` and `_navigateToWebview`, which are referenced in the `build` function. Add the following code after the `build` method:

```
void _navigateToStoryComments(int storyId) {
```

```
   }

   void _navigateToWebview(int storyId) {
   }
```

We will revisit these methods, but for now leave them empty. Now, run the application and you should observe the following:

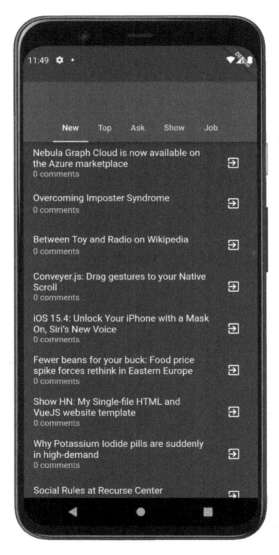

Figure 5.8 – Home view with tabs and stories

Figure 5.8 shows that our application is finally displaying results. Congratulations! As you navigate between tabs and back, you may notice that the tabs must reload with every re-entry. This is less than ideal, as we would prefer the results to be persisted.

Let's update our application to persist the state. Add the `AutomaticKeepAliveClientMixin` mixin to `_StoryListViewState` and then copy the following code before the `initState` method:

```
@override
bool get wantKeepAlive => true;
```

`AutomaticKeepAliveClientMixin` will prevent our widget from being disposed of automatically, as long as `wantKeepAlive` is `true`. Additionally, we need to add `super.build()` as the first line in our `build` method. Now when running the application, you should be rewarded with tabs that persist in their state upon re-entry.

Next, let's change `StoryDetailsView` to display `WebView` which points to the details page URL that the API gives us for each story. Replace the `build` method of `StoryDetailsView` with the following code:

```
return Scaffold(
  appBar: AppBar(),
  body: WebView(
    javascriptMode: JavascriptMode.unrestricted,
    initialUrl:
      'https://news.ycombinator.com/item?id=$storyId',
    gestureRecognizers:
      <Factory<VerticalDragGestureRecognizer>>{
      Factory<VerticalDragGestureRecognizer>(
        () => VerticalDragGestureRecognizer(),
      )
    },
  ),
);
```

In this code snippet, we have swapped the previous body for a `WebView` that points to `https://news.ycombinator.com/item?id=$storyId`, which is the details website for the current story. We also pass in a `VerticalDragGestureRecognizer` to ensure smooth scrolling, since we are using `WebView`. Finally, we allow `WebView` to exercise JavaScript in an unrestricted mode to prevent any rendering errors that might occur.

Next, we would like to add a class to capture the `storyId` that we will pass to the `StoryDetailsView` widget. At the bottom of the file, add the following code:

```
class StoryDetailsViewArguments {
  const StoryDetailsViewArguments(this.storyId);


  final int storyId;
}
```

Here, we have created a class called `StoryDetailsViewArguments` that takes in a single `storyId` parameter, which we will use to build the URL. We can now revisit `story_list_view.dart` and update the `_navigateToStoryComments` function to navigate to our details view using these arguments. Replace the function with the following code:

```
void _navigateToStoryComments(int storyId) {
  Navigator.of(context).pushNamed(
    StoryDetailsView.routeName,
    arguments: StoryDetailsViewArguments(storyId),
  );
}
```

This function will merely redirect to the story details, adding the `arguments` property with the `StoryDetailsViewArguments` that we just created.

As a final step, we need to refactor `app.dart` to make use of the new arguments class. Inside of `MaterialPageRoute.builder` replace the first switch case with the following code:

```
        case StoryDetailsView.routeName:
          assert(routeSettings.arguments != null);
          final arguments =
              routeSettings.arguments as
                StoryDetailsViewArguments;

          return StoryDetailsView(
            storyId: arguments.storyId,
          );
```

The changes that we introduce are as follows:

- Throw an error if no arguments are provided
- Attempt to parse the arguments correctly
- Pass `storyId` from the arguments to the details view

Now, we should be able to run the application and see our details screen when tapping on the `comments` widget:

Figure 5.9 – Story details view

Figure 5.9 shows us the results that we should expect when navigating to the story details view: an empty `AppBar` and a `WebView` that directs us to the details of the story.

Now, let's repeat the process to add a `WebView` for the story's source URL. Create a file called `story_web_view.dart` in the `story` folder. In it, create a `StatefulWidget` called `StoryWebView`. Add the following code to the body of that class:

```
class StoryWebView extends StatefulWidget {
  const StoryWebView({
    Key? key,
    required this.storyId,
  }) : super(key: key);

  final int storyId;

  static const routeName = '/story-source';

  @override
  State<StoryWebView> createState() =>
    _StoryWebViewState();
}
```

Like `StoryDetailsView` this widget takes in a `storyId` parameter and includes `routeName` for use with Navigator. Inside the `_StoryWebViewState` class replace the `build` method with the following:

```
@override
  Widget build(BuildContext context) {
    return ChangeNotifierProvider.value(
      value: _storyController,
      child: Scaffold(
        appBar: AppBar(),
        body: Selector<StoryController, Story?>(
          selector: (context, controller) =>
            controller.selectedStory,
          builder: (context, story, child) {
            if (story == null) {
              return const Center(
                child: CircularProgressIndicator(),
```

```
          );
        }
        if (story.url == null) {
          return const Center(
            child: Text('This story does not have a
                      link'),
          );
        }
        return WebView(
          javascriptMode: JavascriptMode.unrestricted,
          initialUrl: story.url.toString(),
          gestureRecognizers:
            <Factory<VerticalDragGestureRecognizer>>{
            Factory<VerticalDragGestureRecognizer>(
              () => VerticalDragGestureRecognizer(),
            )
          },
        );
      },
    ),
  ),
);
}
```

In this code, we do the following:

- Define a provider for `_storyController`.
- Build a `Scaffold` inside of the provider, selecting the story using `Selector`.
- Display a loading indicator if the story has not loaded.
- Display a message if the story does not have a URL.
- Display a `WebView` if the story has loaded and it also has a URL.

We have not added the logic to retrieve our _storyController or load the story, so we will do that next by adding the following code preceding the build method:

```
late final StoryController _storyController;

@override
void initState() {
  super.initState();
  _storyController = StoryController(context.read())
    ..getStoryById(widget.storyId);
}
```

Now, we just need to round out this page by adding an arguments class to the bottom of the file, similar to StoryDetailsViewArguments.

```
class StoryWebViewArguments {
  StoryWebViewArguments(this.storyId);

  final int storyId;
}
```

Revisit story_list_view.dart and update the _navigateToWebview function to navigate to our source web view using these arguments. Replace the function with the following code:

```
void _navigateToWebview(Story story) {
  Navigator.of(context).pushNamed(
    StoryWebView.routeName,
    arguments: StoryWebViewArguments(story.id),
  );
}
```

This function will merely redirect to the story source view, adding the arguments property with StoryWebViewArguments that we just created.

Finally, we just need to refactor app.dart to make use of the new view and new arguments class. Inside MaterialPageRoute.builder add the following code after the first switch case:

```
            case StoryWebView.routeName:
              assert(routeSettings.arguments != null);
              final arguments =
                  routeSettings.arguments as
```

```
        StoryWebViewArguments;

    return StoryWebView(
      storyId: arguments.storyId,
    );
```

The changes that we introduce are as follows:

- Throw an error if no arguments are provided
- Attempt to parse the arguments correctly
- Pass `storyId` from the arguments to the source view

Now, re-run the application and observe changes, which should lead to the following screen:

Figure 5.10 – Story source view

As you can see from *Figure 5.10*, we now have a view of the story's source URL that matches our details view!

In this section, we finally put our newfound knowledge of `Navigator` to good use, building out our Hacker News application and demonstrating the following:

- How to navigate to pages using arguments
- How to persist page information even as we navigate back and forth between pages.
- How to translate page parameters into data using our previously created `StoryController`

In the next section, we will discover how to build declarative applications with Navigator 2.0 and Google's `GoRouter` library.

Declarative routing with Navigator 2.0

Now that we have learned how to build our Hacker News application using Flutter's legacy navigation APIs, we can refactor our Hacker News Application to use Flutter's declarative Navigator 2.0 APIs. Specifically, we will look at how to use a newer library provided and maintained by the Flutter team called `go_router`. Navigator 2.0 was introduced to provide a more powerful, declarative API that would empower developers who suffered from some of the shortcomings of Navigator 1.0, specifically the following:

- The inability to have more fine-grained control over the navigation stack.
- The inability to properly handle web URLs and deep linking.
- The inability to have state changes easily triggers navigation.

The Navigator 2.0 spec, while more powerful is a more low-level API that is missing many of Navigator 1.0's capabilities, such as `pushReplacement` or `pushNamed`. Fortunately, Google actively maintains a library that abstracts that complexity called `GoRouter`. In this section, we'll learn the following:

- How to set up our application to use Navigator 2.0
- How to define routes in `GoRouter`

When setting up routing using `GoRouter`, there are two classes that you will rely on, `GoRouter` and `GoRoute`. Let's start using them to understand how they work. Create a file called `app_router.dart` in your `src` directory and add the following code:

```
final appRouter = GoRouter(
  routes: [
    GoRoute(
      path: '/',
```

```
      builder: (context, state) => const Material(
        child: Center(
          child: Text('Hello GoRouter'),
        ),
      ),
    ),
  ],
);
```

We have created a simple variable called `appRouter` using the `GoRouter` class, which does the following:

- Define a list of `routes` that will keep track of all the routes in the application
- Create a single `GoRoute` in the `routes` list that uses the / path and will display the centered text **Hello GoRouter** when a user logs in

`GoRouter` is an important class that abstracts the complexity of Navigator 2.0 by automatically creating two classes that it relies on:

- `RouterDelegate`
- `RouteInformationParser`

`GoRoute` declaratively maps URL paths to widgets via the `builder` method. As we will discover later, it additionally handles route nesting along with redirects.

Next, we need to replace our old `MaterialApp` with one that relies on our `appRouter`. Inside of the `app.dart` file, add the dependency on `app_router` to the top of the file and replace `MaterialApp` with the following code:

```
return MaterialApp.router(
  restorationScopeId: 'app',
  localizationsDelegates: const [
    AppLocalizations.delegate,
    GlobalMaterialLocalizations.delegate,
    GlobalWidgetsLocalizations.delegate,
    GlobalCupertinoLocalizations.delegate,
  ],
  supportedLocales: const [
    Locale('en', ''),
  ],
```

```
    onGenerateTitle: (BuildContext context) =>
        AppLocalizations.of(context)!.appTitle,
    theme: ThemeData(),
    darkTheme: ThemeData.dark(),
    themeMode: ThemeMode.dark,
    routeInformationParser:
      appRouter.routeInformationParser,
    routerDelegate: appRouter.routerDelegate,
  );
```

In this code, we have the following:

- Swapped out the `MaterialApp` default constructor for the named `MaterialApp.router` constructor

- Provided the `routeInformationParser` with the one created by our `appRouter`

- Provided the `routerDelegate` with the one created by our `appRouter`

- Removed the `onGenerateRoute` and `home` properties

Now if we re-run our application, it should match the following screen:

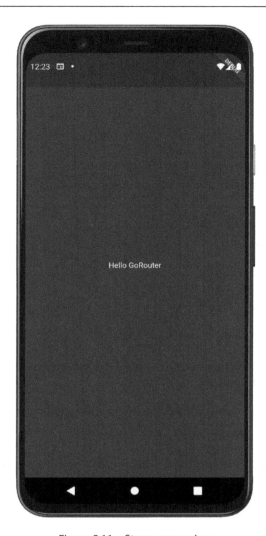

Figure 5.11 – Story source view

From *Figure 5.11*, you should see that the single route that we added previously is displayed. Now, let's extend our example a bit further to demonstrate routing between pages. Replace the routes that we have defined with the following code:

```
GoRoute(
    path: '/',
    builder: (context, state) => Material(
        child: Center(
            child: TextButton(
```

```
            onPressed: () => appRouter.push('/page-two'),
            child: const Text('Go to Page 2'),
          ),
        ),
      ),
    ),
    GoRoute(
      path: '/page-two',
      builder: (context, state) => Material(
        child: Center(
          child: TextButton(
            onPressed: () => appRouter.go('/'),
            child: const Text('Go to Page 1'),
          ),
        ),
      ),
    ),
  ),
```

The code that we have added should define two routes that navigate back and forth between each other using two different methods:

- `appRouter.push` pushes a new route onto the navigation stack.

- `appRouter.go` replaces the entire navigation stack

Now, run the application and observe the following results when clicking back and forth between the buttons:

Figure 5.12 – Story source view

Figure 5.12 shows us the results of running the application and then tapping on the **Go to Page 2** button, which redirects us to the page-two route.

The most powerful feature that we have introduced with these changes is web support. Run the application using flutter run -d chrome and you should observe the following when navigating directly to #/page-two:

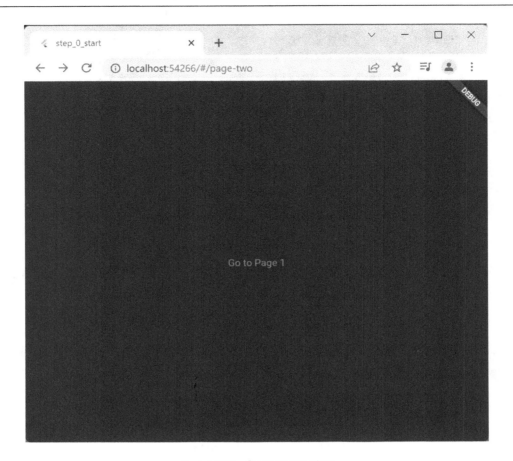

Figure 5.13 – Story source view

From *Figure 5.13*, you can see that the correct page has been displayed without any extra effort on our part. Using Navigator 2.0 in the browser with URLs *just works*.

These capabilities are all the basics that we need to understand about the `go_router` library. Now that we understand how to use Navigator 2.0, we can refactor our Hacker News application to take advantage of more complex scenarios with the popular `go_router` library

Simplifying Navigator 2.0 with GoRouter

Now that we have gotten familiar with how to use `GoRouter` we can begin to refactor the Hacker News application to take advantage of its features. First, we need to swap out our demo routes with the actual routes to our application. Inside of `app_router.dart` replace `GoRouter.routes` with the following:

```
GoRoute(
    path: '/',
    redirect: (_) =>
      '/stories/${StoriesType.newStories.name}',
  ),
  GoRoute(
    path: '/stories/:storyType',
    builder: (context, state) {
      final storyTypeName =
          state.params['storyType'] ??
            StoriesType.newStories.name;
      final storyType =
        StoriesType.values.byName(storyTypeName);

      return HomeView(key: state.pageKey,
                      storyType: storyType);
    },
    routes: [
      GoRoute(
        path: 'story/:id/details',
        name: StoryDetailsView.routeName,
        builder: (context, state) {
          final id = int.parse(state.params['id']!);

          return StoryDetailsView(key: state.pageKey,
                                  storyId: id);
        },
      ),
      GoRoute(
        path: 'story/:id/source',
        name: StoryWebView.routeName,
        builder: (context, state) {
          final id = int.parse(state.params['id']!);

          return StoryWebView(key: state.pageKey,
                              storyId: id);
```

```
      },
    )
  ],
),
```

Make sure to resolve any dependency errors that are reported. This code should not look too dissimilar from our original onGenerateRoute, and does the following:

- Create a / root route that will redirect to the /stories/:storyType route that displays **Home**.

- Create a story/:id/details route that will display the StoryDetailsView widget with the provided storyId.

- Create a story/:id/source that displays the StoryWebView widget with the provided storyId.

The :[parameter] keyword is a special pattern that enables us to include parameters in the route's path. We can then access those parameters from the GoRouterState.params object passed into the builder function.

Next, we need to swap out instances where the application navigates using Navigator for instances using our appRouter. Find the _navigateToStoryComments and _navigateToWebview functions inside of story_list_view.dart and replace them with the following code:

```
void _navigateToStoryComments(int storyId) {
  appRouter.pushNamed(StoryDetailsView.routeName,
      params: {'storyType': widget.storiesType.name,
               'id': '${storyId}'});
}


void _navigateToWebview(int storyId) {
  appRouter.pushNamed(StoryWebView.routeName,
      params: {'storyType': widget.storiesType.name,
               'id': '${storyId}'});
}
```

Now, instead of using Navigator, we are using appRouter and passing the correct story type. If we run the application, we should be met with the result matching our previous example.

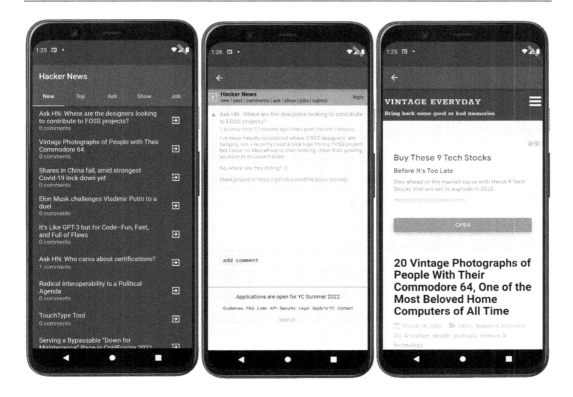

Figure 5.14 – Hacker News app

If the application matches *Figure 5.14*, we should see a full Hacker News application and be able to navigate back and forth between details views and source views.

Now that we have wrapped up our application, let's review what we have learned.

Summary

In this chapter, we learned how to build multi-page applications using Navigator 1.0. Then, we build a more complex Hacker News application to demonstrate how to pass arguments from page to page and use those arguments to display information. Finally, we rewrote the application using `GoRouter` to take advantage of Navigator 2.0's declarative APIs and ability to display pages in the browser, based on their URL.

By learning how to build multi-page applications, you now understand how to provide a more robust experience to users that will work both on mobile devices and in browsers.

In the next chapter, you will learn how to build an application that responds to user input with forms and input fields.

Further reading

- **Navigation and routing**: `https://docs.flutter.dev/development/ui/navigation`
- **GoRouter**: `https://gorouter.dev`

6

Building a Simple
Contact Application with
Forms and Gestures

So far, we have explored how to allow our users to interact with a Flutter application by navigating from page to page, and by tapping on fields to change values or trigger dynamic changes. However, we have not explored how to handle one of the most common use cases of all applications: **data entry**.

You have undoubtedly used an application today that at least required you to input your username and password on a login screen before you could use any of the application's core features. In this chapter, we will learn how to build forms by building a simple Contacts application that allows users to keep track of their network of friends and family members.

In this chapter, we will cover the following topics:

- Setting up the project
- Building forms the manual way
- Building complex forms using `Form`/`FormField` widgets
- Custom form validation

First, we will learn how to build forms manually using `setState` and callback functions. Then, we will explore building our Contacts application using Flutter's `Form` and `FormField` widgets. Finally, we will learn how to validate our form using custom validation.

Technical requirements

Make sure to have your Flutter environment updated to the latest version in the stable channel. Clone our repository and use your favorite IDE to open the Flutter project we've built in `chapter_6/start_0`.

The project that we'll build upon in this chapter can be found on GitHub at `https://github.com/PacktPublishing/Cross-Platform-UIs-with-Flutter/tree/main/chapter_6/start_0`.

The complete source code can be found on GitHub as well: `https://github.com/PacktPublishing/Cross-Platform-UIs-with-Flutter/tree/main/chapter_6/step_3_form_validation`.

Let's set up the project so that we can start to explore how forms work in Flutter applications.

Setting up the project

In this chapter, we will learn how to use Flutter's `form` APIs to build an interactive Contacts application that allows us to create a list of contacts that includes their **First name**, **Last name**, **Phone**, and **Email**.

When we have finished, the resulting application should look as follows:

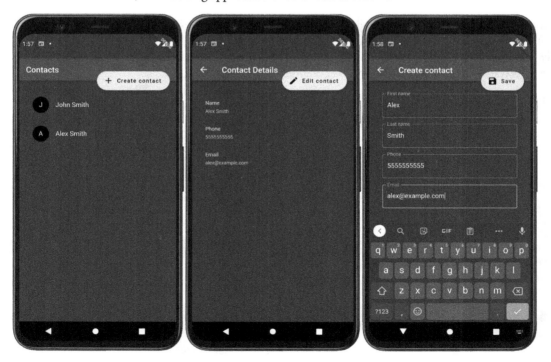

Figure 6.1 – Completed Contacts application

In the preceding figure, we see three different screens:

- A home screen that displays a list of contacts using an avatar of their first initial and their full name

- A details screen that allows us to view each contact's information

- A create/edit screen that allows us to edit existing contacts or create a new one

We will start with an example that already includes the List and Details views in the `contact_list_view.dart` and `contact_details_view.dart` files, respectively. The application uses GoRouter and Provider as dependencies for routing, state management, and dependency injection. It also includes the controllers and APIs to retrieve a list of contacts, add a new contact to the list, or edit a pre-existing contact. After providing an overview of how input fields work in Flutter, we will use Flutter's input widgets to write our Contacts application. Then, we will conclude this chapter by validating the user's input and creating or editing contacts.

After downloading the initial application from GitHub, we can start by moving into its `root` directory and running the `flutter pub get` command from the Terminal. This will install any package dependencies that the application needs. Additionally, opening the project in an IDE will also trigger an installation of the project on initial load and when the `pubspec.yaml` file is updated.

Install the project's dependencies by using the `flutter pub get` command from the project's root. After installing the application's dependencies, execute the `flutter run` command from the same Terminal window and view the project either in your browser or on your currently running emulator:

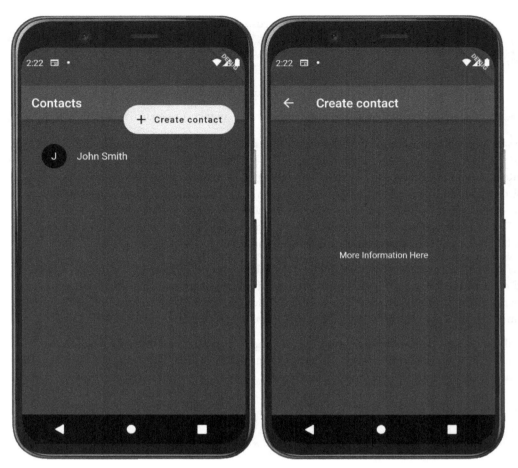

Figure 6.2 – The Contacts List view and Empty Create contact view

On the left-hand side of the preceding figure, observe that we have a pre-populated user, **John Smith**. Upon pressing the **Create contact** button, the application navigates to an empty view, as displayed on the right of the preceding figure. This view, located in `contact_edit_view.dart`, is where we will be working in this chapter.

When viewing the application in your IDE, you should observe the following file structure:

```
lib
|-- src
    |-- data
    |-- localization
```

```
  |-- ui
    |-- contact
  |-- app.dart
|-- main.dart
```

Let's briefly explore the purpose of each top-level folder/class:

- `src`: The source folder contains implementation details of our application that, unless exposed, will be hidden from consumers.

- `data`: This folder includes any reusable business logic for our application. In this case, it includes the Contact model that we will use in our views and our controllers.

- `ui`: This folder includes all the presentational components of our application. This can include controllers and widgets.

- `contact`: This folder includes all of the presentational components associated with contacts.

- `app.dart`: This widget will control whether to display a Material-themed application or a Cupertino-themed application.

- `app_router.dart`: This module includes all of the routing information for our application.

- `main.dart`: This is the entry point of the application.

Now, let's explore the features of the Flutter framework that enable us to build interactive forms.

Building forms the manual way

Flutter has two primary ways of building forms. The first uses `TextField` and similar input field widgets and updates the field values and validation manually. To understand how `TextField` works, let's start by building our contacts form with it.

Open `contact_edit_view.dart` and replace the `Scaffold.body` property with the following code:

```
body: ListView(
  padding: const EdgeInsets.symmetric(
    horizontal: 16,
    vertical: 32,
  ),
  children: const [
    Padding(
      padding: EdgeInsets.symmetric(horizontal:
                              16.0),
```

```
          child: TextField(),
        ),
      ],
    ),
```

Let's examine what this code does:

- It creates a `ListView` that is horizontally padded by `16` pixels and vertically padded by `32` pixels.
- It adds a `TextField` to the ListView's children.

Now, if we reload the application, we should be met with our new field in the Scaffold body:

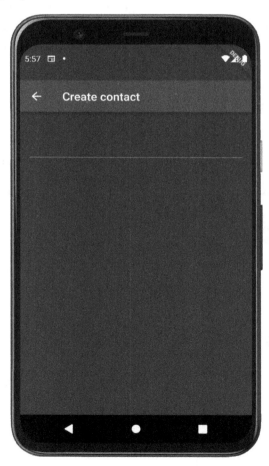

Figure 6.3 – Contact form with an input field

Here, we can see our defined `TextField`. Select the input field and notice that on a mobile device or emulator, the device's soft keyboard displays and hides, depending on whether or not the field is in focus, as shown in *Figure 6.4*:

Figure 6.4 – Contact form with default styling

You may also notice that the default input field is styled with a bottom border that changes colors when the field is in focus and does not provide a label for the field. We can restyle the input field by using `InputDecoration`.

Let's give the field a full border and a label by replacing `TextField` with the following code:

```
TextField(
    decoration: InputDecoration(
```

```
            label: Text('First name'),
            border: OutlineInputBorder(),
        ),
    ),
```

The preceding code adds a `label` called `First name` to the input field and sets the border to `OutlineInputBorder`, which provides the border based on the theme defined in our `MaterialApp`. Now, rerun the application; you should see the following changes:

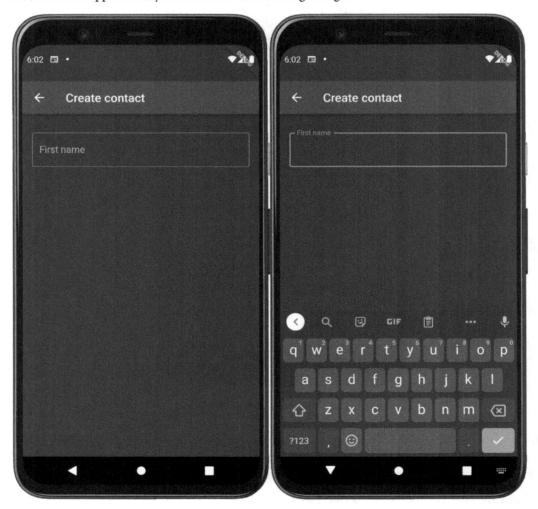

Figure 6.5 – Contact form with outline styling

Notice that in *Figure 6.5*, the unfocused input field renders the label centered inside of the border and that the label transitions to the border when the field is focused. This behavior, again, is the default behavior of TextField in the Material design.

Now, let's learn how to keep track of how the value of TextInput that we have defined changes. The TextField widget accepts an onChanged callback function that is triggered each time the user inputs text. This method could be used in several ways, for example:

- For updating the value of StatefulWidget or Notifier
- For performing validation when the input has changed

Let's use this callback function to validate our newly created field. Update TextField with the following code:

```
TextField(
    decoration: InputDecoration(
        label: const Text('First name'),
        border: const OutlineInputBorder(),
        errorText: _firstNameErrorMessage,
    ),
    onChanged: (value) {
        setState(() {
            _firstNameErrorMessage =
                value.isEmpty ? 'First name is required' :
                    null;
        });
    },
),
```

Additionally, add the _firstNameErrorMessage nullable string variable just before the build method. Now, when you rerun the application, you should be able to add and delete text to/from the input field and observe a transition between unfocused, focused, and error states, as shown in *Figure 6.6*:

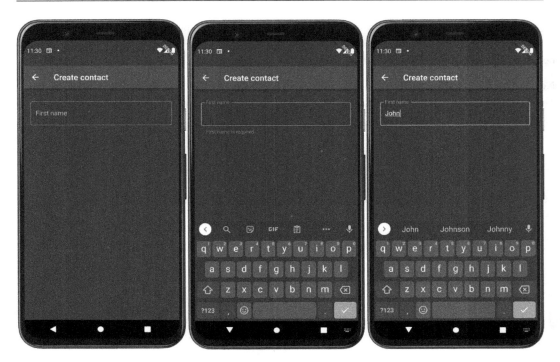

Figure 6.6 – Contact form with error/default styling

We could also defer validation until the field has been submitted by using the `TextField.onSubmitted` callback function property, which has the same signature as the `TextField.onChanged` callback function. To demonstrate, rename `onChanged` to `onSubmitted` and refresh the application.

Upon interacting with `TextField`, you should notice that the states from *Figure 6.6* remain the same, but they are now triggered when the field loses focus via clicking the checkmark on the soft keyboard rather than when the input changes.

`TextField` uses `TextInputAction` to determine whether the field should be placed in a submitted state and uses the default value of `TextInputAction.done` when the input field is not multiline.

Let's see this behavior in action when we have two input fields rather than one. Add the following code after the first `TextField`:

```
const SizedBox(height: 16),
TextField(
    decoration: InputDecoration(
        label: const Text('Last name'),
        border: const OutlineInputBorder(),
```

```
        errorText: _lastNameErrorMessage,
    ),
    onSubmitted: (value) {
        setState(() {
            _lastNameErrorMessage =
                value.isEmpty ? 'Last name is required' :
                null;
        });
    },
),
```

Here, we have added some space between the two fields for aesthetic purposes and then created an input field for Last name. Like _firstNameErrorMessage, add a variable following it called _lastNameErrorMessage. Now, rerun the application; you may notice some odd behavior when using the checkmark key that matches what's shown in *Figure 6.7*:

Figure 6.7 – Contact form with the first and last name input fields

Notice that the first input validates. Upon losing focus, the second input field is not focused correctly and the soft keyboard is dismissed. While `TextField` is smart enough for `TextInputAction. done`, it does not know the surrounding input fields, so it does not know to give focus to the next field in the view.

Therefore, we must do this manually by using another `TextInputAction` – specifically, `TextInputAction.next`. Add `textInputAction: TextInputAction.next` to the first input field's definition and rerun the application. Now, upon interacting with the fields, we should see that validation for each field is triggered when the fields lose focus and a **Next** icon instead of the previous **Done** icon when the first field is in focus. The result should mirror what's shown in *Figure 6.8*:

Figure 6.8 – Contact form with keyboard navigation

Here, we can observe the slightly changed UI of the keyboard and the behavior that we would normally expect.

As a final experiment, let's see how we would perform validation when the user attempts to use another form of interaction, such as clicking a button. To do so, we will use `TextEditingController`, which allows more granular control over the text of `TextField` than a simple callback function. Add the following code at the top of `_ContactEditViewState`:

```
final _firstNameController = TextEditingController();
final _lastNameController = TextEditingController();

@override
void dispose() {
  _firstNameController.dispose();
  _lastNameController.dispose();
  super.dispose();
}
```

Here, we have created an instance of `TextEditingController` for each input field and overridden the class's `dispose` function to ensure that both variables are disposed of correctly. Next, we will remove all instances of `onSubmitted` and change both input fields so that they use their respective controllers. The result of `ListView.children` should resemble the following code:

```
TextField(
    controller: _firstNameController,
    decoration: InputDecoration(
        label: const Text('First name'),
        border: const OutlineInputBorder(),
        errorText: _firstNameErrorMessage,
    ),
    textInputAction: TextInputAction.next,
),
const SizedBox(height: 16),
TextField(
    controller: _lastNameController,
    decoration: InputDecoration(
        label: const Text('Last name'),
        border: const OutlineInputBorder(),
        errorText: _lastNameErrorMessage,
```

```
        ),
    ),
```

Finally, we will need a button to trigger validation. So, let's add `FloatingActionButton` to the scaffold by adding the following code:

```
floatingActionButton: FloatingActionButton.extended(
    onPressed: () {
        setState(() {
            _firstNameErrorMessage =
              _firstNameController.text.isEmpty
                ? 'First name is required'
                : null;
            _lastNameErrorMessage =
              _lastNameController.text.isEmpty
                ? 'Last name is required'
                : null;
        });
    },
    icon: const Icon(Icons.add),
    label: const Text('Create contact'),
),
floatingActionButtonLocation: FloatingActionButtonLocation.
endTop,
```

This code does the following:

- Creates an extended `FloatingActionButton` that allows us to define text and an icon for the button
- Validates both input fields when the button is pressed, using the `TextEditingController.text` property to determine whether the text has been added to the field

Rerun the application and interact with the fields. If you leave a field empty and click on the previously created floating action button, you should be met with correctly validated input fields mirroring the following output:

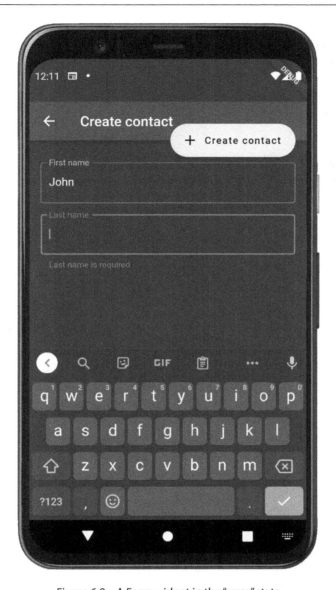

Figure 6.9 – A Form widget in the "error" state

Figure 6.9 shows the results of inputting a value in the first field, leaving the second field blank, and pressing the submission button: an error is displayed on the second field.

Before we move on to the next section, we should cover one last topic: how to set values using `TextEditingController`. The default constructor for `TextEditingController` accepts an optional named parameter.

Let's use it to set our input fields with default values. Update the code that creates the two controllers so that it resembles the following code:

```
final _firstNameController =
  TextEditingController(text: 'John');
final _lastNameController =
  TextEditingController(text: 'Smith');
```

The preceding code defaults the first input field to the value **John** and the second input field to the value **Smith**. Rerun the application; you should see that the input fields have been populated correctly:

Figure 6.10 – A Form widget not in "error" state

Figure 6.10 displays the correctly populated first and last names spread across the two input fields.

Now, we should have a clearer understanding of how to build forms manually. Next, let's learn how to build a form using Flutter's Form and FormField widgets.

Building complex forms with the Form/FormField widgets

Hopefully, the previous section demonstrated that using the vanilla TextField widgets inherently comes with quite a bit of boilerplate since much of the state of each input field must be maintained in the parent widget.

Fortunately, there is a more declarative API that encapsulates most of that boilerplate and allows us to group input fields. Let's explore how to build forms using the Form and FormField widgets.

The Form widget allows you to group multiple input fields to perform operations such as saving, resetting, or validating each input field in unison. The container leverages FormField widgets rather than regular Field widgets, which are its descendants. It allows these operations to be performed by using Form.of with a context whose ancestor is the form, or by using a GlobalKey passed to the form's constructor.

To start, wrap the scaffold in contact_edit_view.dart with a Form widget and a Builder widget:

```
return Form(
    child: Builder(builder: (context) {
        return Scaffold(
            ...
        );
    }),
);
```

We will use the Builder widget to inject a context that can be used to retrieve an instance of its Form ancestor. Next, update the FloatingActionButton.onPressed code so that it resembles the following:

```
if (Form.of(context)!.validate()) {

}
```

Here, we remove the complicated logic from the *Building forms the manual way* section in favor of retrieving the Form ancestor widget and executing its validate function. Upon rerunning the code and interacting with the input fields, you will notice that the validation no longer appears when pressing the **Save** button. This is because we have removed the validation logic.

Fortunately, FormField widgets provide a much simpler and more declarative mechanism to trigger validation: by using the FormField.validator callback function. Let's refactor our ListView. children so that it uses TextFormField widgets instead of TextField widgets:

```
TextFormField(
    controller: _firstNameController,
    textInputAction: TextInputAction.next,
    decoration: const InputDecoration(
        label: Text('First name'),
        border: OutlineInputBorder(),
    ),
    validator: (value) =>
    (value?.isEmpty ?? true) ? 'First name is required' :
      null,
),
const SizedBox(height: 16),
TextFormField(
    controller: _lastNameController,
    textInputAction: TextInputAction.next,
    decoration: const InputDecoration(
        label: Text('Last name'),
        border: OutlineInputBorder(),
    ),
    validator: (value) =>
    (value?.isEmpty ?? true) ? 'Last name is required' : null,
),
```

The only major change to point out is that we now use a validator callback function to return an error message if either input field is empty or null otherwise. Upon calling Form.validate, the form will traverse its FormField ancestors and execute each validator callback function to transition the input field to a valid or invalid state.

Now, if you rerun the application and press the **Save** button without adding any text to the input fields, you should see the following output:

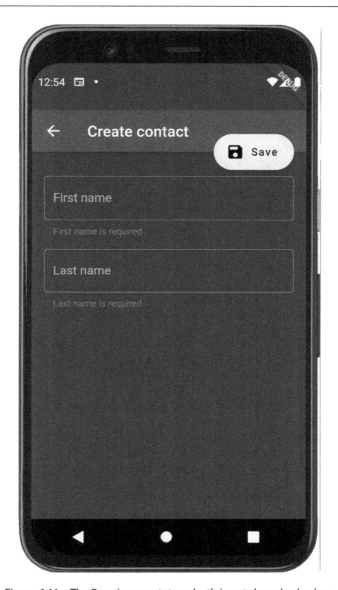

Figure 6.11 – The Form's error state as both inputs have bad values

Figure 6.11 shows that the Contacts application correctly sets the invalid state on both input fields. We can also demonstrate how to reset the form by adding the following code to the end of the `ListView.children` array:

```
ElevatedButton(
    onPressed: () {
```

```
        Form.of(context)!.reset();
    },
    child: const Text('Reset'),
)
```

Now, if we rerun the application and click on the **Reset** button, we should see our form returned to its pristine state:

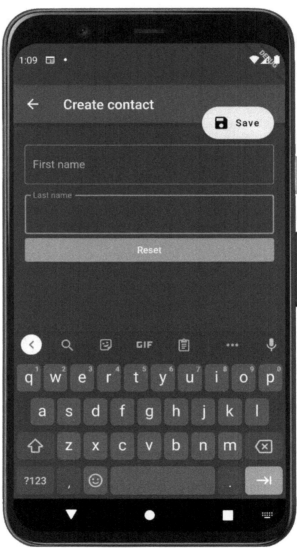

Figure 6.12 – The Form back to its valid state with no more errors

In this section, we learned how to do the following:

- Define a group of input fields using Form and FormField.
- Define a FormField.validator function to validate each input field on submission.
- Use Form.of(context).validate() to validate the form's input fields.
- Use Form.of(context).reset() to reset the form to a pristine state.

In the next section, we will round out our Contacts application by using custom validation.

Custom form validation

Now that we have covered how to use the Form and FormField widgets to reduce some of the boilerplate of handling user input, let's round out our form with new input fields to capture the contact's phone number and email address.

In the contact_edit_view.dart module, add a TextEditingController for the email address and phone number, and then instantiate them in the initState method. The result should resemble the following code:

```
late final TextEditingController _firstNameController;
late final TextEditingController _lastNameController;
late final TextEditingController _emailController;
late final TextEditingController _phoneController;

@override
void initState() {
    super.initState();
    _firstNameController =
       TextEditingController(text: _contact?.firstName);
    _lastNameController =
       TextEditingController(text: _contact?.lastName);
    _emailController =
       TextEditingController(text: _contact?.emailAddress);
    _phoneController =
       TextEditingController(text: _contact?.phoneNumber);
}

@override
void dispose() {
```

```
        _firstNameController.dispose();
        _lastNameController.dispose();
        _emailController.dispose();
        _phoneController.dispose();
        super.dispose();
    }
```

Next, replace the **Reset** button with `FormField` widgets for each input field using the following code:

```
const SizedBox(height: 16),
Padding(
    padding: const EdgeInsets.symmetric(horizontal: 16.0),
    child: TextFormField(
        controller: _phoneController,
        textInputAction: TextInputAction.next,
        decoration: const InputDecoration(
            label: Text('Phone'),
            border: OutlineInputBorder(),
        ),
        validator: _handlePhoneValidation,
    ),
),
const SizedBox(height: 16),
Padding(
    padding: const EdgeInsets.symmetric(horizontal: 16.0),
    child: TextFormField(
        controller: _emailController,
        textInputAction: TextInputAction.done,
        decoration: const InputDecoration(
            label: Text('Email'),
            border: OutlineInputBorder(),
        ),
        validator: _handleEmailValidation,
    ),
),
```

The preceding code creates a form field for the contact's email address and password using the previously created controllers. It also references two validator callback functions that we will now define below the `build` function:

```
String? _handlePhoneValidation(String? value) {
    final isMissing = value?.isEmpty ?? true;

    if (isMissing || !isPhoneNumber(value)) {
        return 'Please provide a valid phone number';
    }

    return null;
}

String? _handleEmailValidation(String? value) {
    final isMissing = value?.isEmpty ?? true;

    if (isMissing || !isEmail(value)) {
        return 'Please provide a valid email address';
    }

    return null;
}
```

The methods in the code sample do the following:

- Return an error if the phone number field is empty or invalid using the `isPhoneNumber` function in `validators.dart`.

- Return an error if the phone number field is empty or invalid using the `isEmail` function in `validators.dart`.

Because the validator function returns a nullable string, we can perform as many custom validations as we would like. The only requirement for validation to display correctly is that an error message is returned. Now, run the application and observe the newly added input fields:

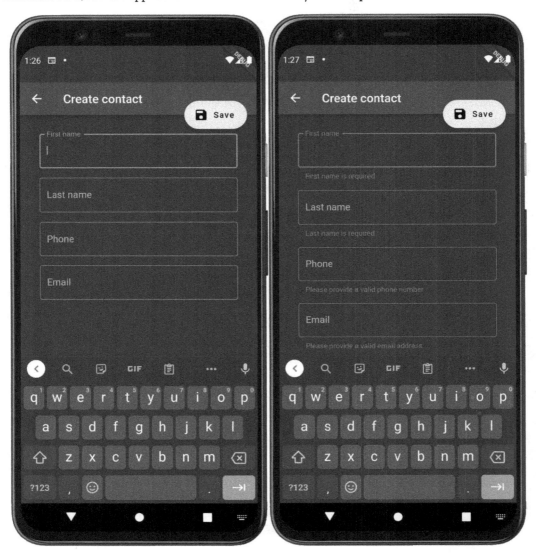

Figure 6.13 – Valid form state on the left, the erroneous one on the right

Figure 6.13 demonstrates the form in a pristine, unsubmitted state and then in an invalid, submitted state after the **Save** button has been clicked. To round out the Contacts application, let's add some code to allow the form to be populated by a previously created contact and create or update the contact when the form is validated. First, add the following code to the beginning of _ContactEditViewState:

```
late final _contact = context
    .read<ContactController?>()
    ?.contacts
    .firstWhereOrNull((contact) => contact.id ==
      widget.id);
```

The preceding code merely retrieves the selected contact from ContactController using the id parameter that was passed into the widget by the router. Next, update the initState method so that it resembles the following code:

```
@override
void initState() {
  super.initState();
  _firstNameController =
      TextEditingController(text: _contact?.firstName);
  _lastNameController =
      TextEditingController(text: _contact?.lastName);
  _emailController =
      TextEditingController(text: _contact?.emailAddress);
  _phoneController =
      TextEditingController(text: _contact?.phoneNumber);
}
```

The preceding code sample sets each controller with the initial value from the previously retrieved contact if it exists. We should be able to refresh the application, navigate to John Smith's contact details view, and then his edit view. The result should resemble the following output:

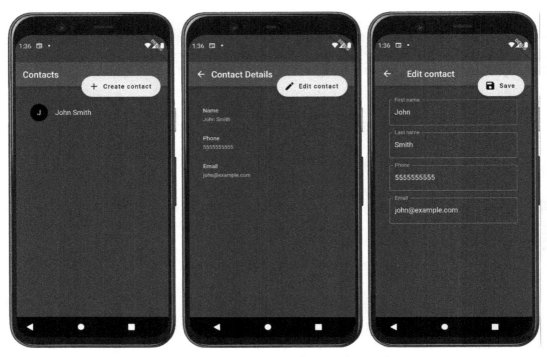

Figure 6.14 – App's navigation flow to edit user data

Figure 6.14 shows the transition that culminates in a pre-populated contact editing form. Finally, let's round out our Contacts application with the logic to create or update a new contact. Replace the `FloatingActionButton.onPressed` function with the following code:

```
onPressed: () {
    ScaffoldFeatureController? scaffoldController;

    if (Form.of(context)!.validate()) {
        final contactController =
          context.read<ContactController>();
        if (_contact == null) {
        } else {
        }
    }
}
Then add the following code to the if block:
        final newContact = Contact(
```

```
            id: contactController.contacts.length + 1,
            firstName: _firstNameController.text,
            lastName: _lastNameController.text,
            emailAddress: _emailController.text,
            phoneNumber: _phoneController.text,
        );
        contactController.addContact(newContact);
        scaffoldController =
            ScaffoldMessenger.of(context).showSnackBar(
            SnackBar(
                content:
                Text('Successfully created
                        ${newContact.firstName}'),
            ),
        );
        await scaffoldController.closed;
        appRouter.goNamed(ContactDetailsView.routeName,
          params: {
            'id': newContact.id.toString(),
        });
```

And finally add the following code to the else block:

```
        final updatedContact = _contact!.copyWith(
            firstName: _firstNameController.text,
            lastName: _lastNameController.text,
            emailAddress: _emailController.text,
            phoneNumber: _phoneController.text,
        );
        contactController.updateContact(updatedContact);
        scaffoldController =
            ScaffoldMessenger.of(context).showSnackBar(
            SnackBar(
                content:
                Text('Successfully updated
                        ${_contact!.firstName}'),
            ),
        );
```

```
                        await scaffoldController.closed;
                        appRouter.pop();
```

The final code sample performs the following logic:

- Initializes a `ScaffoldFeatureController` variable for assignment.

- Retrieves the instance of `ContactController` from the context.

- If the retrieved contact does not exist, creates a new one using the `ContactController.addContact` function and with an `id` incremented from `ContactController.contacts.length`.

- If the retrieved contact exists, updates the contact's values from each input field and updates the contact using the `ContactController.updateContact` function.

- Displays a `SnackBar` message after either operation assigns its controller to the previously initialized `scaffoldController` variable.

- Navigates to the details screen after `SnackBar` has finished displaying.

After rerunning the application, we should be able to create and update contacts using the **Save** button, and be met with a fully functioning Contacts application mirroring what's shown in *Figure 6.15*:

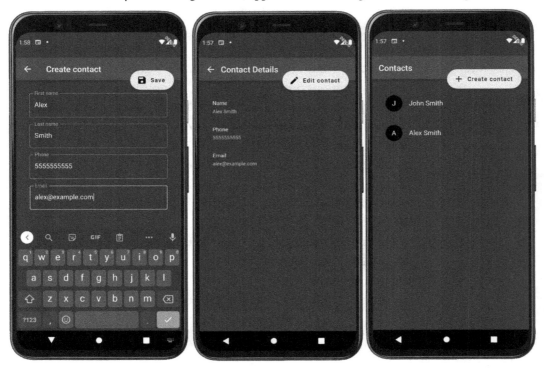

Figure 6.15 – The flow for creating a new contact

Now that we understand how to validate our Contacts form using custom validation and update and create contacts, let's summarize this chapter.

Summary

In this chapter, we learned how to build a form manually using `TextField` widgets. Then, we learned how to reduce some of the boilerplate that we introduced by using the `Form` and `FormField` widgets. Finally, we rounded out our Contacts application by learning how to validate the Contact form using custom validation and updating our contacts when the form has been successfully validated.

You should now have a clear understanding of how to build an application that collects information by using input fields and forms.

In the next chapter, you will learn how to build declarative animations that give your application a unique feel.

Further reading

Forms: `https://docs.flutter.dev/cookbook/forms`.

7

Building an Animated Excuses Application

The true benefits of Flutter shine not just in your ability to build high-fidelity user interfaces but also in allowing you to build high-fidelity experiences. Flutter has the well-stated goal of allowing multiplatform applications to operate at a smooth 60 frames-per-second.

In this chapter, we will learn how to build smooth animations in an application designed to give us random excuses to miss work. First, we will learn about the different types of animations in Flutter. Then, we will build our excuses application to use **explicit animations**. Next, we will refactor the entry of excuses to use **implicit animations**. Finally, we will refactor our Flutter application to take advantage of Flutter's `animations` package.

You will cover the following topics:

- Working with animations in Flutter
- Building implicit animations in Flutter
- Building implicit animations using Flutter's `animations` package

Technical Requirements

Make sure to have your Flutter environment updated to the latest version in the `stable` channel. Clone our repository and use your favorite IDE to open the Flutter project we've built at `chapter_7/start`.

The project that we'll build upon in this chapter can be found on GitHub: `https://github.com/PacktPublishing/Flutter-UI-Projects-for-iOS-Android-and-Web/tree/main/chapter_7/start`.

The complete source code can be found on GitHub as well: `https://github.com/PacktPublishing/Flutter-UI-Projects-for-iOS-Android-and-Web/tree/main/chapter_7/step_3`.

Let's set up the project so that we can start to explore how animations work in a Flutter application.

Setting up the project

In this chapter, we will learn how to use Flutter's animation APIs to build an interactive Excuses application that allows us to randomly select an excuse to skip work.

When we have finished, the resulting application should look like *Figure 7.1*:

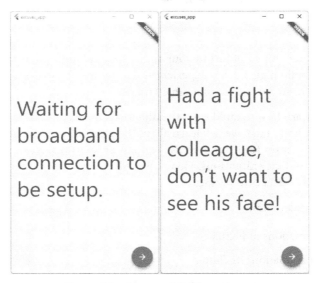

Figure 7.1 – A page of the Excuses app

In *Figure 7.1*, we see the same screen with two different texts that will animate in the following ways:

- An exit animation for the progress indicator when loading the excuses from the server
- An entry animation when the excuses finish loading from the server
- An exit animation for the old excuse loading the next excuse
- An entry animation when loading the next excuse

We will start with an example that already includes the Excuse view, in files named `excuses_view.dart`. The application uses `Equatable`, `Dio`, and `Provider` as dependencies for equality, HTTP requests, and dependency injection. It also includes the API facade to retrieve a list of excuses.

After downloading the initial application from GitHub, we can start by moving into its root directory and running the `flutter pub get` command from the terminal. This will install any package dependencies that the application needs. Additionally, opening the project in an IDE will also trigger the installation of the project on initial load and when `pubspec.yaml` is updated.

Install the project's dependencies by using the `flutter pub get` command from the project's root, and after installing the application's dependencies, execute the `flutter run` command from the same Terminal window. View the project either in your browser or on your currently running emulator. Here are two images of the Excuses app running on Windows:

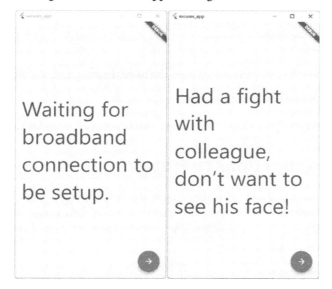

Figure 7.2 – The Excuses app running in Windows

In the first image of *Figure 7.2*, observe that the application already pulls back instances of excuses. Upon pressing the **Next** icon button the application snaps to the next excuse without any transition.

When viewing the application in your IDE, you should observe the following file structure:

```
lib
|-- src:
    |-- data
    |-- localization
    |-- ui
        |-- excuse
    |-- app.dart
|-- main.dart
```

Let's briefly explore the purpose of each top-level folder/class:

- `src`: The source folder contains implementation details of our application that, unless exposed, will be hidden from consumers.

- `data`: This folder includes any reusable business logic for our application. In this case, it includes the `Excuse` model and the `ExcuseFacade` service that we will use in our views and our controllers.

- `ui`: This is the folder that includes all of the presentational components of our application. This can include controllers and widgets.

- `excuse`: This is the folder that includes all of the presentational components associated with excuses.

- `app.dart`: This is the widget that initializes the basic theming and routing of the application.

- `main.dart`: This is the entry point of the application.

Now let's explore the features of the Flutter framework that enable us to build interactive animations.

Working with animations in Flutter

In Flutter there are two ways to build animations: explicitly and implicitly. Explicit animations are a set of APIs and controls that tell the Flutter framework how to rebuild the widget tree while modifying widget properties to create a seamless animation effect. While this effect is more manual and low-level than implicit animations, it is more powerful.

Implicit animations enable you to defer the management of animations to the Flutter framework. Instead of manually controlling the animation, it can be triggered by updating widget properties. Changes to a widget's property will trigger the animation from the previous value to the newly specified one.

To understand how implicit animations work, let's start by building our contacts form with it.

Open up `excuses_view.dart`. Replace the `_ExcusesViewContentState` class definition with the following code:

```
class _ExcusesViewContentState extends State<_
ExcusesViewContent>
    with TickerProviderStateMixin {
```

In this code snippet, we have extended the `State` object with `TickerProviderStateMixin`. In Flutter, animations are controlled at a high level by two special classes: `AnimationController` and `Animation`. `Animation` is a special type of `Listenable` that adds the ability to track its status and the current value as it transitions. `AnimationController` is a special class that allows you to play/pause/reverse animations or set specific values.

When using `AnimationController` in a `StatefulWidget State` object, you must use the `TickerProviderStateMixin` or `TickerProviderStateMixin` class to properly notify `AnimationController` and its container whenever a frame triggers.

As you may have guessed from the previous paragraph, we will be working closely with `AnimationController` and `Animation` in this section. Now let's add two instances of `AnimationController` and `Animation` to control the exit transition of the loading indicator and the entry animation of the first excuse. Inside `_ExcusesViewContentState`, add the following code:

```
late Animation<double> _loadingAnimation;
late AnimationController _loadingAnimationController;
late Animation<double> _excusesAnimation;
late AnimationController _excusesAnimationController;

@override
void initState() {
super.initState();
  _loadingAnimationController = AnimationController(
      duration:
        const Duration(milliseconds: 500), vsync: this);
  _loadingAnimation =
      Tween<double>(begin: 1,
        end: 0).animate(_loadingAnimationController);
  _excusesAnimationController = AnimationController(
      duration:
        const Duration(milliseconds: 500), vsync: this);
  _excusesAnimation =
      Tween<double>(begin: 0,
        end: 1).animate(_excusesAnimationController);

}
```

In this code, we do the following:

- Instantiate instances of `Animation` and `AnimationController`.

- Initialize the instances of both `AnimationController` with `duration` and `vsync` properties. The `vsync` property ensures that the animation only runs while the screen is displayed.

- Initialize the instances of Animation with `Tween.animate` constructors, passing in `begin` and `end` properties.

When initializing the instances of AnimationController, we use a duration of 500 milliseconds to define the lifetime of the animation, and we set the vsync property to configure TickerProvider for each AnimationController. When initializing the instances of Animation, we use tweens - stateless objects that take begin and end properties and handle the mapping between those values using AnimationController. Because we want to animate the hiding of the loading indicator and the showing of the excuses, we flip the begin/end values, using 1.0/0.0 for _loadingAnimation and 0.0/1.0 for _excusesAnimation.

The _loadingAnimation will transition from a beginning value of 1.0 to 0.0 for its opacity. The _excusesAnimation will transition from a beginning value of 0.0 to an end value of 1.0 for its opacity.

Next, we want to determine when the animations will be triggered using the didUpdateWidget lifecycle function. Add the following code to _ExcusesViewContentState after the initState function.

```
@override
void didUpdateWidget(covariant _
  ExcusesViewContent oldWidget) {
  super.didUpdateWidget(oldWidget);

  if (widget.excuses != null) {
    _loadingAnimationController
        .forward()
        .then((value) =>
        _excusesAnimationController.forward());
  } else {
    _excusesAnimationController
        .reverse()
        .then((value) =>
        _loadingAnimationController.reverse());
  }
}

@override
void dispose() {
  _loadingAnimationController.dispose();
  _excusesAnimationController.dispose();
  super.dispose();
}
```

The previous code does the following:

- Plays the loading animation controller and then plays the excuses animation controller if the excuses have been loaded.

- Reverses the excuses animation controller and then reverses the loading animation controller if the excuses have not been loaded.

- Disposes of the animation controllers when the `dispose` lifecycle function of the `State` object is triggered.

Finally, we need to make our animations come to life. We will do this by using the `AnimationBuilder` class. If you remember examples of `AnimationBuilder` from previous chapters, the widget uses an instance of `Listenable` to rebuild its children.

Replace the child of `Scaffold | SafeArea | Padding` with the following code:

```
child: Stack(
  children: [
    AnimatedBuilder(
      animation: _loadingAnimation,
      builder: (context, child) {
        return Opacity(
          opacity: _loadingAnimation.value,
          child: child,
        );
      },
      child: const Center(
        child: CircularProgressIndicator(),
      ),
    ),
    if (widget.excuses != null)
      AnimatedBuilder(
        animation: _excusesAnimation,
        builder: (context, child) {
          return Opacity(
            opacity: _excusesAnimation.value,
            child: child,
          );
        },
        child: ExcusePageView(
```

```
                          excuses: widget.excuses!,
                          currentExcuse: currentPage,
                    ),
                 )
             ],
          ),
```

This code does the following:

- Uses a `Stack` to overlay the loading indicator and the excuses on top of each other.
- Lays out two `AnimationBuilder` that depend on `_loadingAnimation` and `_excusesAnimation` respectively.
- Sets the builder of `AnimatedBuilder` of the progress indicator to animate the widget's opacity from a visible state to a transparent state.
- Sets the builder of `AnimatedBuilder` of `ExcusePageView` to animate the widget's opacity from a visible state to a transparent state.

Now upon refreshing the application, notice that the choppy transition between the progress indicator and the initial excuse has been replaced with a smooth transition where the progress indicator's opacity eases out before the first excuse's opacity eases in, as in *Figure 7.3*:

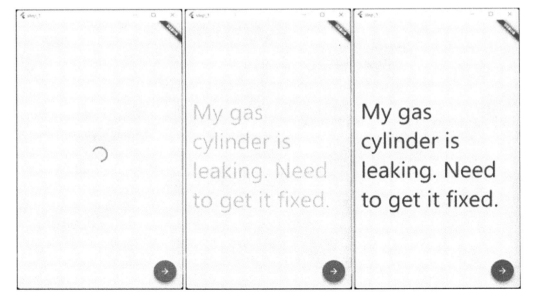

Figure 7.3 – An excuse fading into visibility

Now we should have a clearer understanding of how to build explicit animations by extending `StatefulWidget` with the `TickerProviderStateMixin` class, then initializing instances of `Animation` and `AnimationController` classes that specify what values will transition and how, and finally bringing the explicit animation together using `AnimationBuilder` to subscribe to the changes of the `Animation` class.

Next, let's learn how to animate the transition between excuses using implicit animations.

Building implicit animations in Flutter

Hopefully, the previous section demonstrated that explicit animations are a powerful tool for triggering custom, staggered animations, but also can require a bit of boilerplate.

Fortunately, as with every layer in Flutter's architecture, animations can be controlled implicitly, or declaratively, with a set of widgets purpose-built to encapsulate some of the complexities of animations. Let's explore how to animate the transition between excuses using implicit animations.

In Flutter, you can defer the management of animations to the framework by using a collection of *implicitly animated widgets*. Rather than dealing with `Animation` and `AnimationController` objects directly, these widgets allow you to animate a widget's properties while encapsulating the use of those objects.

`AnimatedOpacity` is one such widget, which accepts a `duration` and animates over that duration when the `opacity` value changes, using the old value and the new value as the begin and end respectively. We could accomplish the task from the previous section using the following code inside of `Scaffold | SafeArea | Padding`.

```
child: Stack(
  children: [
    AnimatedOpacity(
      duration: Duration(milliseconds: 500),
      opacity: widget.excuses != null ? 1 : 0,
      onEnd: () => setState(() {
        _showExcuses = true;
      }),
      child: const Center(
        child: CircularProgressIndicator(),
      ),
    ),
    if (widget.excuses != null)
      AnimatedOpacity(
```

```
        duration: Duration(milliseconds: 500),
        opacity: _showExcuses ? 1 : 0,
        child: ExcusePageView(
          excuses: widget.excuses!,
          currentExcuse: currentPage,
        ),
      )
    ],
  ),
```

This code sample does the following

- Uses `AnimatedOpacity` to hide the progress indicator when the excuses load.

- Sets the `_showExcuses` property to `true` when the progress indicator's animation ends.

- Uses `AnimatedOpacity` to show the initial excuse when `_showExcuses` is `true`.

Basically, the state of the `excuses` and `_showExcuses` properties declaratively triggers the transitions. There are even widgets that allow you to animate between an unknown number of children, such as `AnimatedSwitcher`. Let's use `AnimatedSwitcher` to dynamically swap out excuses when the **Next** icon button is clicked. Navigate to `excuse_page_view.dart`. Replace the `build` method of `ExcusePageView` with the following code:

```
var selectedExcuse = excuses[currentExcuse];

return AnimatedSwitcher(
  duration: const Duration(milliseconds: 500),
  transitionBuilder:
    (Widget child, Animation<double> animation) {
    return ScaleTransition(scale: animation,
                              child: child);
  },
  child: ExcuseCard(
    key: ValueKey(selectedExcuse.id),
    excuse: selectedExcuse,
  ),
);
```

This code does the following:

- Wraps ExcuseCard with an AnimatedSwitcher
- Sets the duration property to **500 milliseconds**
- Changes the transition from the default to ScaleTransition, setting the scale property to the animation

The AnimatedSwitcher will switch between a new widget and the previously set widget using the transition provided by the transition builder. Now run the code and tap next a few times. You should notice a transition where one excuse fades out while another fades in, as in *Figure 7.4*:

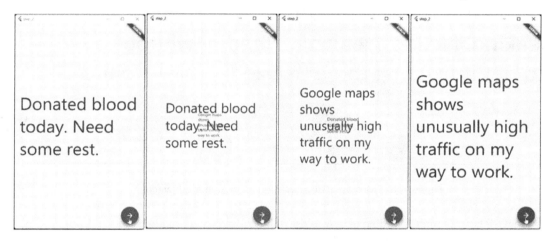

Figure 7.4 – Some animation frames in the Excuses app

In *Figure 7.4*, we see the transition from the first excuse to the second excuse. The previous excuse starts to scale down as the next excuse scales up.

If no transition is provided, AnimatedSwitcher with a cross-fade animation by default. Remove transitionBuilder so that the code for AnimatedSwitcher matches the following:

```
return AnimatedSwitcher(
  duration: const Duration(milliseconds: 500),
  child: ExcuseCard(
    key: ValueKey(selectedExcuse.id),
    excuse: selectedExcuse,
  ),
);
```

In this code sample, we removed `transitionBuilder` and should see a transition where one excuse fades out while another fades in, as in *Figure 7.5*.

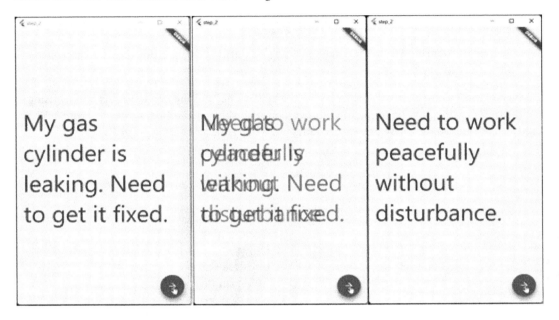

Figure 7.5 – The cross-fade animation of an AnimatedSwitcher

Now that we have an understanding of how to use different implicit animations, we can learn how to utilize even more high-level widgets to elicit more robust implicit animations.

Building implicit animations using Flutter's animations package

Now that we have learned how to leverage implicit and explicit animations, let's refactor our code to take advantage of Flutter's `animations` package, which contains pre-canned animations for common use cases from the Material motion system spec.

First, install the `animations` package by running `flutter pub add animations` in the project's root directory in the terminal.

Next, swap out `AnimatedSwitcher` in `excuse_page_view.dart` with `PageTransitionSwitcher` from the package. The `PageTransitionSwitcher` is a variation of the `AnimatedSwitcher` that allows for separate transitions to be used for the entry and exit animations. Replace the `build` method of `ExcusePageView` with the following code:

```
final selectedExcuse = excuses[currentExcuse];

return PageTransitionSwitcher(
    duration: const Duration(milliseconds: 500),
    transitionBuilder:
     (child, animation, secondAnimation) {
        return SlideTransition(
            position: Tween<Offset>(
                begin: Offset.zero,
                end: const Offset(1.5, 0.0),
            ).animate(secondAnimation),
            child: FadeTransition(
                opacity: Tween<double>(
                    begin: 0.0,
                    end: 1.0,
                ).animate(animation),
                child: child,
            ),
        );
    },
    child: ExcuseCard(
        key: ValueKey(selectedExcuse.id),
        excuse: selectedExcuse,
    ),
);
```

In this example, we first use `SlideTransition` with the `secondAnimation` to cause the old child to exit by sliding out while `FadeTransition` is used to animate the new child's entry by fading in. Like `AnimatedSwitcher`, `PageTransitionSwitcher` needs a way to determine whether it should transition. For this reason, we set a `ValueKey` on `ExcuseCard` using the ID of the current excuse.

Now rerun the application and press the **Next** icon button a few times. You should see a very clean animation matching *Figure 7.6*:

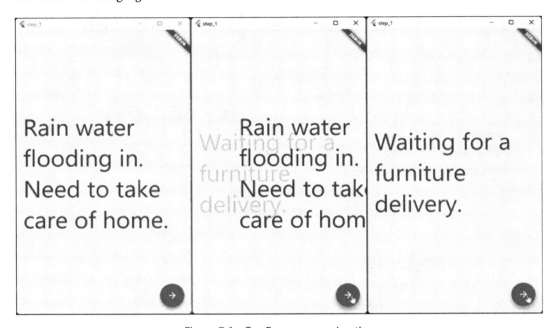

Figure 7.6 – Our Excuse app animating

As previously described, we watch the first excuse slide out in *Figure 7.6* as the next excuse fades into view.

The PageTransitionSwitcher allows us to create an infinite amount of entry/exit transitions. To simulate this, we will swap the transitions while adding ScaleTransition that also utilizes secondAnimation. The result should match the following code:

```
return FadeTransition(
    opacity: Tween<double>(
        begin: 1.0,
        end: 0.0,
    ).animate(secondAnimation),
    child: ScaleTransition(
        scale: Tween<double>(
            begin: 1.0,
            end: 0.0,
```

```
    ).animate(secondAnimation),
    child: SlideTransition(
        position: Tween<Offset>(
            end: Offset.zero,
            begin: const Offset(1.5, 0.0),
        ).animate(animation),
        child: child,
    ),
  ),
);
```

Upon running this code sample, the old excuse should fade out and scale down in size as the new excuse slides into view. Press the **Next** icon button a few times. Your result should match *Figure 7.7*:

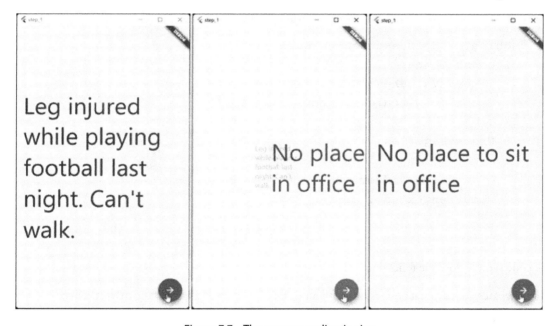

Figure 7.7 – The excuses scaling in size

With just a few code changes, we have created an even more robust transition! Let's round off our Excuses application by replacing the animation logic in excuses_view to take advantage of PageTransitionSwitcher from the animations package.

First, delete all instances of Animation and AnimationController, including the initState, didUpdateWidget, and dispose lifecycle functions. Then replace the child of Scaffold | SafeArea | Padding with the following code:

```
child: PageTransitionSwitcher(
    duration: const Duration(milliseconds: 1000),
    transitionBuilder: (child, animation, secondAnimation) {
        return SharedAxisTransition(
            child: child,
            animation: animation,
            secondaryAnimation: secondAnimation,
            transitionType:
                SharedAxisTransitionType.horizontal,
        );
    },
    child: widget.excuses == null
    ? const Center(
        key: ValueKey('progress'),
        child: CircularProgressIndicator(),
    )
    : ExcusePageView(
        key: const ValueKey('excuses'),
        excuses: widget.excuses!,
        currentExcuse: currentPage,
    ),
),
```

The code sample does the following:

- Replaces the explicit animations with an instance of PageTransitionSwitcher
- Uses the SharedAxisTransition widget from the animations package for transitionBuilder
- Adds a ValueKey to the progress indicator and the excuse page view to trigger the transition

We have not discussed SharedAxisTransition. It is a built-in transition from the animations package that combines fade/slide transitions across the horizontal or vertical axes or scale transitions depending on the value of SharedAxisTransitionType used. In this case, we have utilized SharedAxisTransitionType.horizontal, which will cause a fade and slide-in transition for the entry animation and a fade and slide-out transition for the exit animation.

Upon running this code sample, you should see the **Next** icon button a few times. Your result should match *Figure 7.8*:

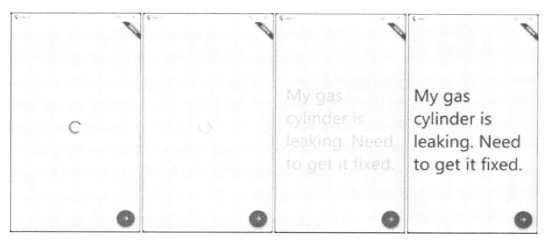

Figure 7.8 – The full Excuses app

Figure 7.8 should demonstrate how you have reduced the size of your code base while also adding more complex animations for the entry/exit transition of the progress indicator and excuse page view.

Now that we have learned how to leverage the `animations` package to build complex and robust transitions and implicit animations, let's recap what we learned in this chapter.

Summary

In this chapter, we learned about the different types of animations in Flutter. Then, we learned how to leverage explicit animations in our Excuses application. We refactored the application to utilize implicit animations, driven by property changes. Finally, we refactored our Flutter application to leverage Flutter's `animations` package.

Throughout this chapter, you have learned about the trade-offs of implicit animations and explicit animations. The important takeaway is implicit animations are just abstractions over explicit animations that encapsulate common use cases. As your application grows and you build custom animations, you might discover common use cases unique to your application. That is a perfect opportunity to create an implicit animation to keep your code reuse as simple as possible.

You now have a good understanding of how to build high-fidelity user experiences using your own custom animations.

In the next chapter, you will learn how to build a responsive application using Flutter and Supabase.

Further reading

- **Explicit animations**: `https://docs.flutter.dev/codelabs/explicit-animations`

- **Implicit animations**: `https://docs.flutter.dev/codelabs/implicit-animations`

- **Animations package**: `https://pub.dev/packages/animations`

- **AnimatedBuilder**: `https://api.flutter.dev/flutter/widgets/AnimatedBuilder-class.html`

8

Build an Adaptive, Responsive Note-Taking Application with Flutter and Dart Frog

Flutter is a multiplatform framework, which means that we are not just limited to one platform or a handful of screen sizes. To truly maximize the potential of Flutter, our application must adapt to being used on different platforms and respond to being used in different screen sizes.

In this chapter, we will learn how to leverage layout widgets such as `MediaQuery` and `ConstrainedBox` (just to name a few) to build a responsive and adaptive application. We will create a note-taking application like Google Keep, where users can create and edit notes and view them as a list, and demonstrate responsive and adaptive design by having the application's UI change depending on the platform being either mobile or desktop.

So, first we will learn about the differences between responsive and adaptive layouts. Then, we will enhance our application to be responsive. Next, we will enhance our application to be adaptive. Finally, we add a backend API for our application and refactor the application to use the backend API rather than local data.

In this chapter, we'll cover the following topics:

- An introduction to responsiveness and adaptiveness
- Making an app responsive
- Making an application adaptive
- Creating REST endpoints for the Notes application

Technical requirements

Make sure to have your Flutter environment updated to the latest version in the `stable` channel. Clone our repository and use your favorite IDE to open the Flutter project we've built at `chapter_8/start`.

The project that we'll build upon in this chapter can be found on GitHub: `https://github.com/PacktPublishing/Cross-Platform-UIs-with-Flutter/tree/main/chapter_8/start`

The complete source code can be found on GitHub as well: `https://github.com/PacktPublishing/Cross-Platform-UIs-with-Flutter/tree/main/chapter_8/step_5`.

Let's set up the project so that we can start to explore how to build responsive Flutter applications.

Setting up the project

In this chapter, we will learn how to use Flutter's layout widgets and APIs to build a responsive, adaptive Notes application that allows us to create, edit, and delete notes.

When finished, the resulting application on your mobile should mirror *Figure 8.1*:

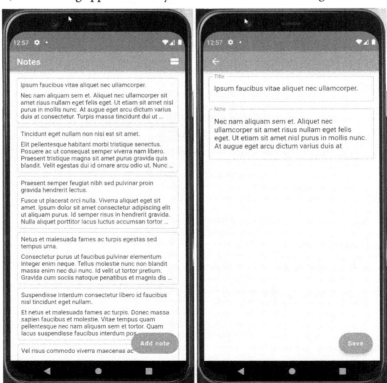

Figure 8.1 – The Notes application running on a mobile platform

And if you view the application on your desktop, it should mirror *Figure 8.2*:

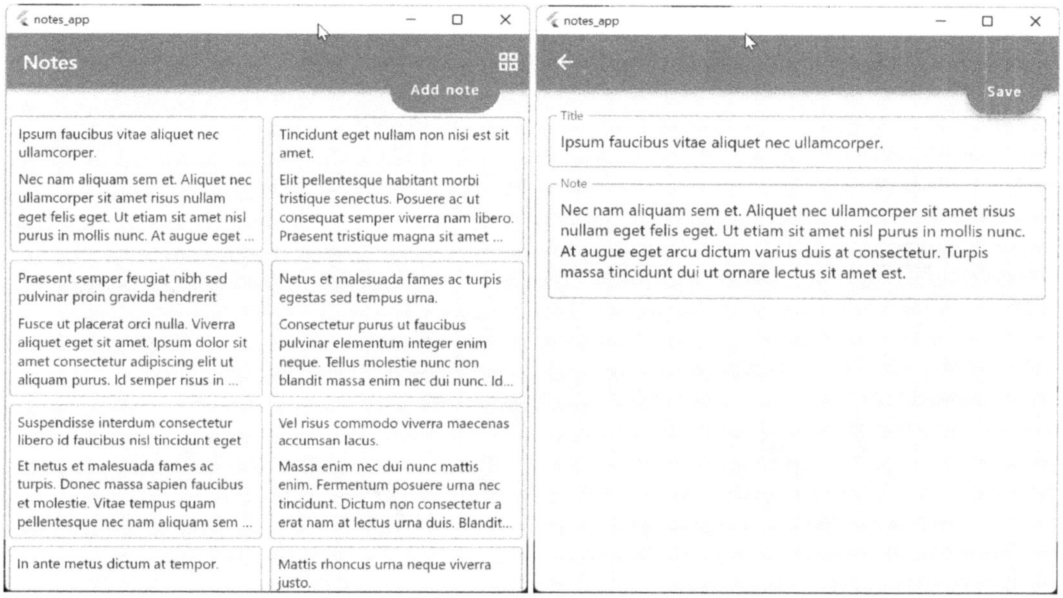

Figure 8.2 – Notes application running on a desktop platform

In *Figure 8.1*, we see the notes application when running on a mobile platform. In *Figure 8.2*, we see the same application running on a desktop platform. Notice the following differences:

- The mobile list view renders as a vertical single-column list, while the desktop list view renders as a multi-column grid.

- `FloatingActionButton` floats in a mobile view and is docked in `AppBar` in the desktop view.

- Dimensions such as padding are changed depending on the layout of the application.

We will start with an example that already includes the views to view, create, and edit notes. The application uses `Equatable`, `Dio`, and `Riverpod` as dependencies for equality, HTTP requests, and caching and dependency injection respectively. It keeps track of a list of notes locally. After an overview of responsive and adaptive design, we will use responsive mechanisms to allow the application to respond to different screen sizes. Then we will use adaptive mechanisms to allow the application to adapt to different platforms. Finally, we will conclude the chapter by replacing the local data with a backend API.

After downloading the initial application from GitHub, we can start by moving into its root directory and running the `flutter pub get` command from the terminal. This will install any package dependencies that the application needs. Additionally, opening the project in an IDE will also trigger an installation of the project on initial load and when the `pubspec.yaml` file is updated.

Install the project's dependencies by using the `flutter pub get` command from the project's root, and after installing the application's dependencies, execute the `flutter run` command from the same terminal window and view the project either in your browser or on your currently running emulator.

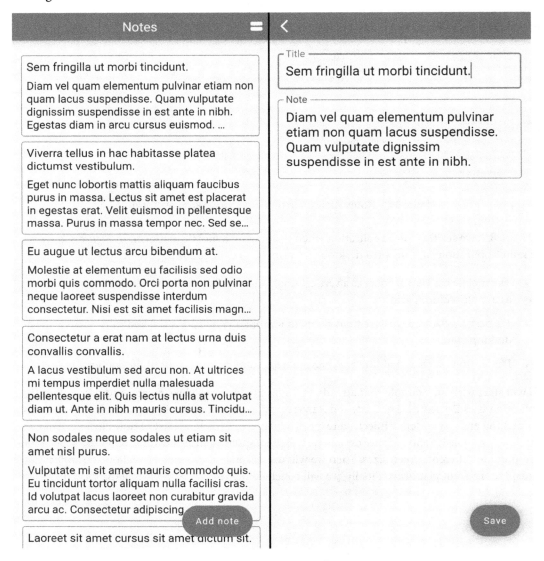

Figure 8.3 – The starting application

In the left-hand screenshot in *Figure 8.3*, note that the application renders a list of notes on the initial screen. Upon pressing the **Add note** button the application redirects to a form that can be used to create a note. Tapping on an existing note from the initial screen will redirect to the same form but to edit the existing note.

When viewing the application in your IDE, you should observe the following file structure:

```
notes_app
    lib
    |-- src:
        |-- data
        |-- localization
        |-- ui
            |-- notes
        |-- app.dart
    |-- main.dart
packages
    notes_common
        lib
        |-- notes_common.dart
```

Let's briefly explore the purpose of each top-level folder/class:

* notes_app: The project for the Notes frontend application.

    * src: The source folder contains implementation details of our application that, unless exposed, will be hidden from consumers.

    * data: This folder includes any reusable business logic for our application. In this case, it includes the NoteFacade service that we will use in our views and our controllers.

    * ui: The folder that includes all the presentational components of our application. This can include controllers and widgets.

        * note: The folder that includes all the presentational components, controllers, and providers associated with notes.

    * app.dart: The widget that initializes the basic theming and routing of the application.

    * main.dart: The entry point of the application.

- `packages/notes_common`: The project for any reusable components across the frontend and backend.

For now, `notes_common` will only be used in the frontend application, but we will be adding a backend in a later section.

Hopefully, you should notice some similar patterns to previous chapters, such as defining providers for dependency and state management, creating a class to handle application routing, and project organization.

Now let's explore the features of the Flutter framework that enable us to build responsive and adaptive applications.

Introduction to responsiveness and adaptiveness

In Flutter you will commonly hear the terms **adaptive** and **responsive** when referring to building applications. While these terms are both related to layout, they have very different meanings:

- **Responsive design** refers to adjusting the layout of the application for the available screen size.
- **Adaptive design** refers to adjusting the behavior, layout, and even the UI of the application for the platform or device type in use, such as mobile, desktop, or web.

An application can be responsive without being adaptive, or adaptive without being responsive. Alternatively, an application can be neither. We have all opened a fair share of applications that, regardless of the device, still look and behave like mobile applications. The starter version of our Notes application does just that.

Run `flutter run -d {platform}` from the `notes_app` folder, substituting `{platform}` for the desktop platform that you are developing on. Even though you are running in a desktop environment, note that the application looks like a giant phone application.

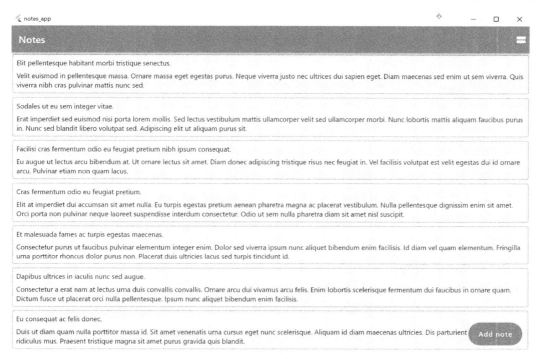

Figure 8.4 – The Notes list view in a desktop environment

Figure 8.4 shows a list view with all the pre-generated notes in giant tiles. The call-to-action button to add a new note is in the lower right corner of the application, an odd position for a desktop environment, and the button in `AppBar` has an unusually large splash radius.

The application is currently being displayed in a **mobile-first layout**. As the term suggests, mobile-first denotes that the *application is built for mobile devices first*.

The note creation page is not much better:

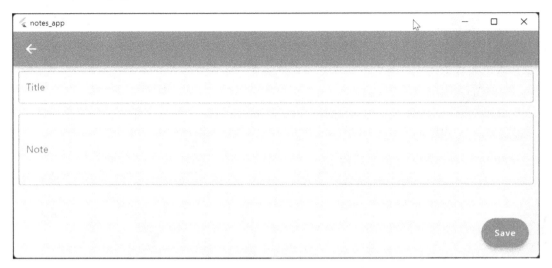

Figure 8.5 – The Notes creation page viewed in a desktop environment

Like the notes cards in *Figure 8.4*, the form in *Figure 8.5* grows to fill the entire application window. While this design is perfectly reasonable on a mobile device, it can create bad user experiences in a browser or desktop environment.

Instead, we would like our application to **adapt** to make the most of screen real estate and the environment in which it is running. This is where the **responsive** and **adaptive** APIs and widgets of Flutter come in handy; they allow you to create applications that can *adapt* to the device's environment, screen size, and orientation.

Now we should have a clearer understanding of the differences and uses of adaptive and responsive design. Next, let's learn to make the Notes application responsive using Flutter APIs and widgets.

Making an app responsive

Hopefully, the previous section demonstrated the importance of building applications that can look nice no matter what environment they are running in, especially when using a multiplatform framework such as Flutter. Now let's explore how to make the Notes frontend application responsive, allowing us to adjust the layout of the application for the available screen size.

Take a look at the Notes application running on the desktop. If you resize the window, you'll notice that regardless of the size of the window, the grid view remains two-column. While this is fine for a mobile or tablet device, on a large screen we are wasting screen real estate. Instead, we will use the `ScreenType` enum defined in `utils/screen_type.dart` to decide how many rows to display. Open that file and examine its contents:

```
enum ScreenType {
  desktop._(minWidth: 901),
  tablet._(minWidth: 601, maxWidth: 900),
  handset._(maxWidth: 600);

  const ScreenType._({
    this.minWidth,
    this.maxWidth,
  });

  factory ScreenType.fromSize(double deviceWidth) {
    if (deviceWidth > ScreenType.tablet.maxWidth!) return
      ScreenType.desktop;
    if (deviceWidth > ScreenType.handset.minWidth!) return
      ScreenType.tablet;

    return ScreenType.handset;
  }

  final int? minWidth;
  final int? maxWidth;
}
```

The previous code does the following:

- Defines three screen types with optional minimum and maximum widths: desktop, tablet, and handset
- Defines a factory function to compute the screen type from the device width

Now let's use these values to determine how many grid columns to display. Open up the `notes_list_view.dart` file and append the following code to the block that returns `StaggeredGrid`:

```
if (layout == Layout.grid) {
    final screenSize =
        MediaQuery.of(context).size.shortestSide;
    final crossAxisCount = min(
        (screenSize / ScreenType.handset.minWidth!).floor(),
        4);

    //...
}
```

In this code, we grab the screen size of the application and then compute the grid column count by dividing it by the minimum width of a handset, choosing the minimum of that value or 4. This will ensure that our grid column count never grows beyond 4. Refresh the app and observe that resizing the window changes the row count, mirroring *Figure 8.6*.

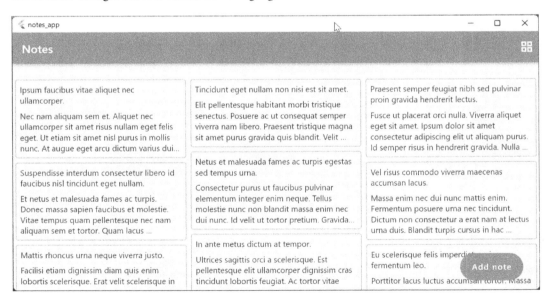

Figure 8.6 – A non-responsive Notes list view

Figure 8.6 shows the application with three rows when the screen width is larger than 900 pixels.

Also note that once the application reaches a size where four columns can be supported, the columns just continue to grow to match the window size. We would prefer the windows to max out at a certain width. Fortunately, we can use the `ConstrainedBox` widget to accomplish this by passing it a `maxWidth` constraint. Wrap `StaggeredGrid` with `ConstrainedBox` using the following code:

```
return ConstrainedBox(
    constraints: const BoxConstraints(maxWidth: 1200),
    child: StaggeredGrid.count(
        //...
    ),
);
```

In this code, we give `ConstrainedBox` a `maxWidth` constraint of 1200 pixels. This is the equivalent of telling the `StaggeredGrid` to stop growing in width once it has reached 1200 pixels. Rerun the application, and you should now see that at a certain width, the `StaggeredGrid` remains centered, mirroring *Figure 8.7*.

Figure 8.7 – A responsive Notes list view

As displayed in *Figure 8.7*, no matter how large the application grows, the grid will not increase in size.

Now that we have an understanding of how to use different implicit animations we can learn how to utilize even more high-level widgets to elicit more robust implicit animations.

Making an app adaptive

Now that we have learned how to build responsive applications, adjusting the layout of the Notes application for the available screen size, let's switch focus to making our application adaptive.

Recall that **adaptive design** refers to adjusting the behavior, layout, and even the UI of the application to the platform or device type in use, such as mobile, desktop, or web. For examples of adaptive widgets that already exist, look no further than Flutter's Material framework. A tooltip, for example, has very different behavior depending on the platform that it is rendered.

A tooltip is a widget that enhances another visual element with additional information while maintaining a minimal interface. In environments that support a mouse as an input, the Material tooltip's default behavior is to display itself when its target is hovered and dismiss itself when that same target exits a hover state.

In touch-enabled environments or when a mouse is not present, the default behavior of the tooltip is to display itself when its target is in a long-pressed state and dismiss itself when that target exits a long-pressed state.

In Flutter, we can make our own adaptive components by checking the platform that the application is currently running on. Open the `device.dart` file in the `notes_app` project. Observe the following code:

```
import 'dart:io';
import 'package:flutter/foundation.dart';


bool get isMobileDevice => !kIsWeb && (Platform.isIOS ||
Platform.isAndroid);
bool get isDesktopDevice =>
    !kIsWeb && (Platform.isMacOS || Platform.isWindows ||
                Platform.isLinux);
bool get isMobileDeviceOrWeb => kIsWeb || isMobileDevice;
bool get isDesktopDeviceOrWeb => kIsWeb || isDesktopDevice;
```

Here we have created several helper variables that we can use throughout the application:

- `isMobileDevice` will check whether the application is running on a mobile device such as iOS or Android.

- `isDesktopDevice` will check whether the application is running on a desktop device such as Windows.

- `isMobileDeviceOrWeb` will check whether the application is running on a mobile device or in the browser.

- `isDesktopDeviceOrWeb` will check whether the application is running on a desktop device or in the browser.

Let's start by changing the default orientation of the Notes list page to be a vertical list when the application is running on a mobile device and a staggered grid for all other platforms. Change `layoutProviderProvider` to the following code:

```
final layoutProviderProvider = StateProvider<Layout>((ref) {
  return isMobileDevice ? Layout.list : Layout.grid;
});
```

Here we are using the `isMobileDevice` global variable to alter the default layout. Upon refreshing the application, we should see a different layout on each platform, matching *Figure 8.8*.

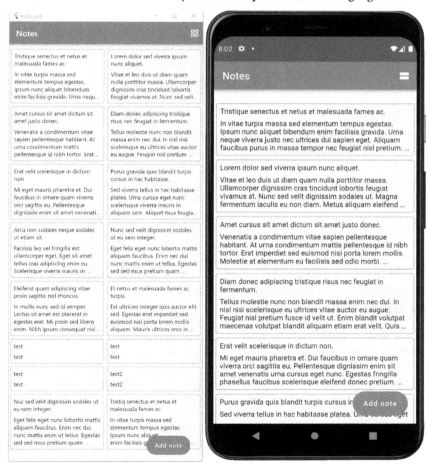

Figure 8.8 – An adaptive notes list view

Next, let's change the location of `FloatingActionButton` to be docked in `AppBar` when the application is running on a desktop device or in the browser. Inside of `notes_list_view.dart` add the following code to the `Scaffold` widget:

```
return Scaffold(
  //...
  floatingActionButton: Padding(
    padding: isDesktopDeviceOrWeb
        ? const EdgeInsets.only(right: 12)
        : EdgeInsets.zero,
    child: FloatingActionButton.extended(
      elevation: 0,
      onPressed: () {
        appRouter.pushNamed(
          NoteDetailsView.routeNameCreate);
      },
      label: const Text('Add note'),
    ),
  ),
  floatingActionButtonLocation: isMobileDeviceOrWeb
      ? FloatingActionButtonLocation.endFloat
      : FloatingActionButtonLocation.endTop,
  //...
);
```

This code does the following:

- Adds `FloatingActionButton` that will navigate to the note creation view when pressed
- Changes the positioning of `FloatingActionButton` based on whether the app is running on a mobile device or the web

Now run the code targeting desktop environments and you should observe how the application displays differently in each environment.

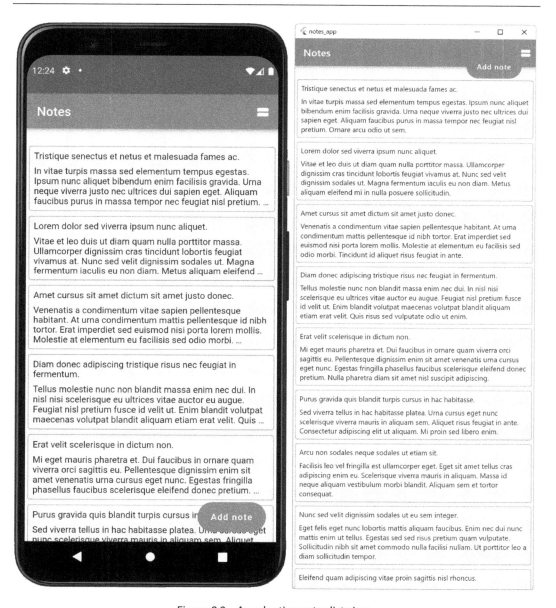

Figure 8.9 – An adaptive notes list view

Figure 8.9 displays a mobile version of the application where the `FloatingActionButton` is displayed in the bottom right corner and a desktop version where the `FloatingActionButton` is docked in `AppBar`. If you hover over the menu action button in a desktop environment, you can see that it has an unnecessarily large splash radius. Let's tweak this behavior to be exclusive to the mobile platform by updating `AppBar` with the following code:

```
appBar: AppBar(
  title: const Text('Notes'),
  actions: [
    Consumer(builder: (context, ref, _) {
      final layout =
          ref.watch(layoutProviderProvider);
      final iconData =
          layout == Layout.grid ? Icons.grid_view :
              Icons.view_stream;

      return IconButton(
        splashRadius: isDesktopDeviceOrWeb ? 16 :
            null,
        icon: Icon(iconData),
        onPressed: () {
          final newLayout =
              layout == Layout.grid ? Layout.list :
                  Layout.grid;
          ref.read(layoutProviderProvider.notifier)
            .state = newLayout;
        },
      );
    }),
  ],
),
```

With the preceding code, we will display a much smaller splash radius in a desktop environment or web environment where the user is likely to be be using a mouse for input.

Next, let's make similar enhancements to the form for creating notes. Open up `note_form.dart` and add the following code to `Scaffold`:

```
return Scaffold(
  //...
  floatingActionButtonLocation: isMobileDeviceOrWeb
      ? FloatingActionButtonLocation.endFloat
      : FloatingActionButtonLocation.endTop,
  //...
);
```

This code will also change the position of `FloatingActionButton`. Additionally, we should add some spacing to account for the button when it is docked in the `AppBar`. Add the following code to the `ListView` widget:

```
padding: isDesktopDeviceOrWeb
    ? const EdgeInsets.only(left: 12,
                                right: 12, top: 26)
    : const EdgeInsets.all(12),
```

Upon re-running the application and navigating to the Notes **Create** or **Edit** page, we should see results matching *Figure 8.10*.

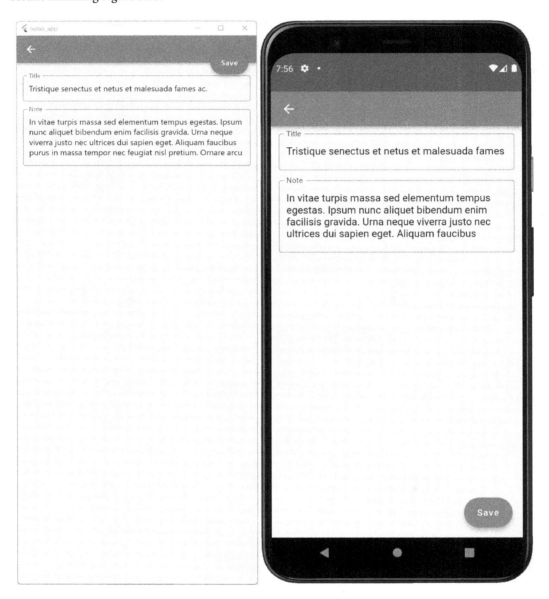

Figure 8.10 – An adaptive Notes edit/create view

Finally, let's restrict the size of the form using a `ConstrainedBox` and by adding alignment using the `Align` widget. Wrap `ListView` with the following code:

```
child: Align(
    alignment: Alignment.topCenter,
    child: ConstrainedBox(
      constraints: const BoxConstraints(
        maxHeight: 420,
        maxWidth: 720,
      ),
      child: ListView(
        //...
      ),
    ),
),
```

The previous code does the following:

- Aligns the form at the top and center of the body of `Scaffold`
- Constrains the width of the form to not increase past 720 pixels
- Constrains the height of the form to not increase past 420 pixels

Upon rerunning the application, we should now see a much better design for desktop and web, matching *Figure 8.11*.

Figure 8.11 – An adaptive Notes create/edit view

Now that we have learned how to leverage Flutter's adaptive mechanisms to create an application that adapts to its environment, let's wrap up this chapter by creating a backend for the Notes application.

Creating REST endpoints for the Notes application

Up until this chapter, we have mainly relied on free third-party APIs when building out our Flutter applications. Unfortunately, as developers, we won't always be able to rely on open APIs or a backend team to build the services we need. Fortunately, the Dart language is just as capable of building APIs as it is for building Flutter applications. In this section, we will wrap up our Notes application by building our own backend API.

To build our API, we will be using `dart_frog`, a fast and minimalistic framework that is heavily inspired by Node.js and Next.js and allows developers to build backends in Dart. Start by installing `dart_frog` globally using the following command:

```
dart pub global activate dart_frog_cli
```

Once the command is finished running, we are ready to create our project. Run the following command at the root of the project, a level above the application:

```
dart_frog create notes_api
```

After running this command, we should see a new folder called `notes_api` that contains our backend code. When viewing the application in your IDE, you should observe the following file structure:

```
routes
|-- index.dart
test
|-- routes
    |-- index_test.dart
```

Let's briefly explore the purpose of each top-level folder/class:

- `routes`: The source folder contains routes for the API.

  - `index.dart`: A pre-generated API route.

- `test`: The folder that includes all of the tests for the API.

In Dart Frog, a route is made by creating a `.dart` file in the `routes` directory that exports a route handler `onRequest` function. Open the `index.dart` file and you should see the following code:

```dart
import 'package:dart_frog/dart_frog.dart';

Response onRequest(RequestContext context) {
  return Response(body: 'Welcome to Dart Frog!');
}
```

This example route merely returns a welcome message when queried. Now let's start the API by running `dart_frog dev` in the folder. After the script is finished running, navigate to `http://localhost:8080` in the browser. We should see a welcome message once the page is finished loading, mirroring *Figure 8.12*.

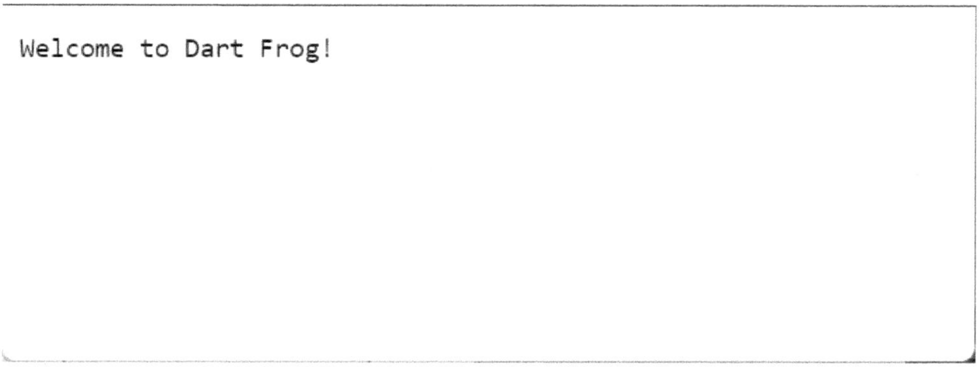

```
Welcome to Dart Frog!
```

Figure 8.12 – The initial API response

Before creating routes for our API, let's first migrate our mock data to the backend. Create a `data` folder and add a `notes.dart` file to it with the following code:

```dart
import 'package:faker/faker.dart';
import 'package:notes_common/notes_common.dart';

List<Note> notes = List.generate(
  10,
  (index) => Note(
    id: index.toString(),
```

```
      title: _faker.lorem.sentence(),
      content: _faker.lorem.sentences(5).join(' '),
    ),
  );


  final _faker = Faker();
```

In the previous code, we are generating 10 notes with random data. We will use this notes list in the routes that we are about to create.

Next, let's add a route to return our notes. Create a notes directory in the routes folder and add an index.dart file with the following code:

```
import 'package:dart_frog/dart_frog.dart';
import 'package:notes_common/notes_common.dart';

import '../../data/notes.dart';

Future<Response> onRequest(RequestContext context) async {
  switch (context.request.method) {
    case HttpMethod.get:
      return _handleGet(context);
    case HttpMethod.post:
      return _handlePost(context);
    // ignore: no_default_cases
    default:
      return Response(statusCode: 404);
  }
}
```

This code does the following:

- Forwards all GET requests to the _handleGet function

- Forwards all POST requests to the _handlePost function

- Returns a 404 error in any other case, denoting that the route could not be handled

Now add the missing functions to the file with the following code:

```
Response _handleGet(RequestContext context) {
  return Response.json(
    body: notes.map((e) => e.toMap()).toList(),
  );
}


Future<Response> _handlePost(RequestContext context) async {
  final body = await context.request.json();
  final note = Note.fromMap(body).copyWith(id:
      (notes.length + 1).toString());

  notes = [...notes, note];

  return Response.json(
    body: note.toMap(),
  );
}
```

In this code, the _handleGet function will return all our notes in the response after converting them to JSON. The _handlePost function will retrieve the JSON payload from the request, create a new note and add it to our list of notes, and then return the newly created note in the response, also converting it to JSON.

We are still missing the ability to delete and edit notes. In both cases, we would like to include the unique identifier of the note in the URL of the API request. Fortunately, Dart Frog supports path parameters by using dynamic routes, using the convention of creating the filename with brackets around the parameter. Create a file in the notes folder called [id].dart. Now the onRequest route handler will be provided with an extra path parameter for the note's unique identifier. Open the file and add the following code:

```
import 'package:dart_frog/dart_frog.dart';
import 'package:notes_common/notes_common.dart';

import '../../data/notes.dart';
```

```
Future<Response> onRequest(RequestContext context, String id)
async {
  switch (context.request.method) {
    case HttpMethod.put:
      return _handlePut(context, id);
    case HttpMethod.delete:
      return _handleDelete(context, id);
    // ignore: no_default_cases
    default:
      return Response(statusCode: 404);
  }
}
```

This code should look very similar to the first route handler that we defined:

- Forwards all PUT requests to the _handlePut function

- Forwards all DELETE requests to the _handleDelete function

- Returns a 404 in any other case

Now add the missing functions to the file with the following code:

```
Future<Response> _handlePut(RequestContext context, String id)
async {
  final body = await context.request.json();
  final note = Note.fromMap(body).copyWith(id: id);
  notes = notes.map((e) => e.id == note.id ? note :
      e).toList();

  return Response.json(
    body: note.toMap(),
  );
}

Response _handleDelete(RequestContext context, String id) {
  notes = notes.where((note) => note.id != id).toList();
```

```
    return Response(statusCode: 204);
}
```

The _handlePut and _handleDelete functions will handle editing and removing notes respectively. Now we can start to refactor our application to correctly use our API. Inside of the notes_service.dart file, enhance the NotesService constructor with the following code to accept dio as a parameter to be able to make HTTP requests:

```
class NotesService {
  NotesService(this.dio);

  final Dio dio;
  //...
}
```

Next, update the provider to pass in an instance of dio with the following code:

```
final notesServiceProvider = Provider<NotesService>((ref) {
  final baseUrl =
      isMobileDevice ? 'http://10.0.2.2:8080' :
          'http://localhost:8080';
  final options = BaseOptions(baseUrl: baseUrl);

  return NotesService(Dio(options));
});
```

Note that when creating an instance of Dio we set the baseUrl parameter based on whether or not the application is targeting a mobile platform. This is done because the emulator does not have knowledge of localhost.

Next, we will refactor the create and getAll functions to call the API with the following code:

```
class NotesService {
  //...
  Future<List<Note>> getAll() async {
    final response = await
        dio.get<List<dynamic>>('/notes');
```

```
    final notes = response.data
           ?.cast<Map<String, dynamic>>()
           .map(Note.fromMap)
           .toList() ??
       [];

  return notes;
}

Future<void> create({
  required String title,
  required String content,
}) async {
  await dio.post('/notes', data: {
    'title': title,
    'content': content,
  });
}
//...
}
```

This code does the following:

- Inside of the getAll function, it creates a GET request to retrieve the notes from the API.
- Inside of the create function, it creates a POST request to the API to create a new note.

Finally, we just need to refactor the update and delete functions with the following code:

```
class NotesService {
  //...
  Future<void> update({
    required String id,
    required String title,
    required String content,
  }) async {
    await dio.put('/notes/$id', data: {
```

```
      'title': title,
      'content': content,
    });
  }


  Future<void> delete({
    required String id,
  }) async {
    await dio.delete('/notes/$id');
  }
  //...
}
```

This code does the following:

- Inside of the `update` function, it creates a PUT request to update an existing note, using the unique identifier as a path parameter.
- Inside of the `delete` function, it creates a DELETE request to delete an existing note, using the unique identifier as a path parameter.

Upon running the application, we should see no changes in our screens and now be able now to retrieve, create, update, and delete notes from the backend API that we have created.

Now that we have completed our backend and frontend updates, we know how to efficiently build a full stack Flutter application. Let's review what we have learned in this chapter.

Summary

In this chapter, we learned adaptive and responsive design, how to use responsive mechanisms to change the layout of the application to accommodate different screen sizes, and how to use adaptive mechanisms to alter the behavior and display of the application. We started with an application that used mobile-first design and enhanced it to respond to different screen sizes. Then we enhanced the application further to alter its behavior depending on its target platform. Finally, we moved our hardcoded data to a backend server that is written in `dart_frog`.

You should now have a good understanding of how to build truly immersive applications that look good no matter where they are run.

In the next chapter, you will learn how to build write tests to ensure that you deliver high-quality applications to your users.

Further reading

- **Creating responsive and adaptive apps:** `https://docs.flutter.dev/development/ui/layout/adaptive-responsive`
- **Building adaptive apps:** `https://docs.flutter.dev/development/ui/layout/building-adaptive-apps`

9
Writing Tests and Setting Up GitHub Actions

Taking care of rebuilding widgets only when needed, using the latest version of each package, and following Flutter's best coding practices will surely point you toward success. However, there's more you need to take into account if you want your product to be of the highest possible quality. In this chapter, we're going to provide an overview of how testing works in Flutter and which tools should you use to create rock-solid apps.

Other than writing tests, another important part of setting up is having a CI pipeline that automatically checks your code's health, executes tests, compares results, and shows reports to the developer. For this reason, we will dive into GitHub to set up actions, PR templates, and much more.

After that, we're going to take an existing project and improve it with unit, widget, and golden tests. Then, we will take care of the GitHub repository and make sure that all of our tests and analysis rules are systematically checked by the repository.

In this chapter, we are going to cover the following topics:

- Strategies for good unit and widget testing in Flutter
- Writing golden tests
- Setting up GitHub actions

Let's get started!

Technical requirements

Make sure that your Flutter environment has been updated to the latest version in the `stable` channel. Since we're going to be working on the application that we made in *Chapter 1, Building a Counter App with History Tracking to Establish Fundamentals*, please clone our repository and use your favorite IDE to open the Flutter project we built at `chapter_1/with_inherited_widget`.

As always, the complete code for this chapter can be found in this book's GitHub repository at `https://github.com/PacktPublishing/Cross-Platform-UIs-with-Flutter/tree/main/chapter_9`. Note that GitHub's settings, including workflows for using GitHub actions, will be located at the root of the repository, so you'll find the `.github` folder at `https://github.com/PacktPublishing/Cross-Platform-UIs-with-Flutter`.

Setting up the project

First of all, you need to create a top-level folder called `test`; this is where we're going to spend most of our time. Then, you need to add a new dependency to the `pubspec.yaml` file since we're going to use it in the next section. Once you've done this, your `dev_dependencies` should be as follows (if it's the case, make sure that you update them to their latest versions):

```
dev_dependencies:
  dart_code_metrics: ^4.8.1
  flutter_lints: ^1.0.4

  flutter_test:
    sdk: flutter

  golden_toolkit: ^0.12.0
```

Finally, we need to create a useful widget that wraps a common configuration that's required by all the widgets we're going to test. Since the app is based on a material design, our class is going to wrap a `MaterialApp` and a `Scaffold` widget. Place this widget inside the `test` folder and name it whatever you want; for example, `mock_wrapper.dart`:

```
class MockWrapper extends StatelessWidget {
  final Widget child;

  const MockWrapper({
    Key? key,
    required this.child,
  }) : super(key: key);
```

```
  @override
  Widget build(BuildContext context) {
    return MaterialApp(
      debugShowCheckedModeBanner: false,
      home: Scaffold(
        body: child,
      ),
    );
  }
}
```

With that, we can start writing some tests!

Writing tests for the Flutter app

For the sake of consistency, we generally recommend keeping the same file organization in both the lib and test folders. This will guarantee a better mental order and allow you to quickly search for the tests when you need to work on them.

To get started, let's create the main_test.dart file in the test folder and ensure it has the following contents:

```
import 'package:chapter_1/main.dart' as app_main;

void main() {
  group('Testing the root widget', () {
    testWidgets(        "Making sure that 'main()' doesn't
throw",
        (tester) async {
        var throws = false;

        try {
          app_main.main();
        } catch (_) {
          throws = true;
        }

        expect(throws, isFalse);
```

```
      },
    );

    testWidgets('Making sure that the root widget is
      rendered', (tester) async {
      await tester.pumpWidget(const
        app_main.EnhancedCounterApp());

      expect(find.byType(CounterAppBody), findsOneWidget);
      expect(find.byType(MaterialApp), findsOneWidget);
    });
  });
}
```

In the first test, we ensure that the app can correctly start without raising any exceptions, even if there isn't much value for it in this specific project since we don't have any startup initialization tasks. If we did some preliminary work before calling runApp() instead, such as setting up a crash report tool or some platform-specific API calls, then this test would become very important.

In the second test, we simply check that the root widget instantiates some fundamental widgets we always expect to have. To be more precise, this is a **smoke test** because its purpose is to reveal simple failures that would compromise the health of the app.

> **Warning**
> Don't underestimate the value of smoke tests! Their purpose is to make sure that some fundamental pieces of a component are there and in the right place. Anything is destined to collapse without some solid basis to stand on, including software, so ensuring that the foundations are there is always a good practice to follow!

Now, let's create the counter_app folder inside test to emulate the structure we have inside lib. Before moving on, we need to create the test/flutter_test_config.dart file:

```
Future<void> testExecutable(FutureOr<void> Function()
  testMain) async {
  // Loading fonts for golden tests
  setUpAll(() async {
    await loadAppFonts();
  });
```

```
  // Tests' main body
  await testMain();
}
```

In the *Writing widget and golden tests* section, you will understand why this function is so important to define. Make sure that you use the same name we've given because Flutter looks for this specific file, if any, before setting up the testing engine.

Now, let's move on to the next testing phase!

Writing unit tests

We're testing classes while following the order in which we encounter them in `lib`, so let's create a `model` folder that contains a new file called `counter_test.dart`. Since the `Counter` class is not a Flutter widget, we're going to write a unit test rather than a widget test:

```
test('Making sure that the class is correctly initialized', ()
{
  final counter = Counter();

  expect(counter.history.length, isZero);
  expect(counter.counter, isZero);
});
```

Here, we've made sure that the state variables we care about are initialized to zero – this is a smoke test. Now, we can test the behavior of the class to verify that the counter and the history list are updated correctly whenever we call `increase()` or `decrease()`:

```
test('Making sure that values can be increased', () {
  final counter = Counter()
    ..increase()
    ..increase();

  expect(counter.history, orderedEquals(<int>[1, 2]));
  expect(counter.counter, equals(2));
});

test('Making sure that values can be decreased', () {
  final counter = Counter()
    ..increase()
```

```
    ..increase()
    ..decrease();

  expect(counter.history, orderedEquals(<int>[1, 2]));
  expect(counter.counter, equals(1));
});
```

Since our `Counter` class mixes with a `ChangeNotifier`, we also need to make sure that both `increase()` and `decrease()` also notify listeners when called. We can add a listener with a custom counter to check whether `notifyListeners()` calls have been executed or not:

```
test('Making sure that listeners are notified', () {
  var listenerCount = 0;

  final counter = Counter()
    ..addListener(() => ++listenerCount);

  // Increase
  counter.increase();
  expect(listenerCount, equals(1));

  // Decrease
  counter.decrease();
  expect(listenerCount, equals(2));
});
```

At this point, we've covered all the behaviors of the `Counter` class with unit tests. Since the model folder doesn't contain anything else to test, let's create a new folder called `test/counter_app/widgets` so that we can start writing widgets and tests.

Writing widget and golden tests

Writing **golden tests** is part of the widget testing process and in this section, we're going to learn how to write them. The only widget we have to test in the `widgets` folder is `CounterAppBody`, so let's write a smoke test to ensure that basic widgets are rendered in the tree:

```
testWidgets('Making sure that the widget is rendered', (tester)
async {
  await tester.pumpWidget(const MockWrapper(
    child: CounterAppBody(),
```

```
  ));

  expect(find.byType(CounterAppBody), findsOneWidget);
  expect(find.byType(AppTitle), findsOneWidget);
  expect(find.byType(HistoryWidget), findsOneWidget);
  expect(find.byType(ElevatedButton), findsNWidgets(2));
});
```

This is not enough, of course! We need to make sure that this widget behaves as expected. When tapping one of the two buttons, the Text widget at the center (representing the current count) must update alongside the history list.

Let's use tester.tap() to simulate a finger (or a cursor) pressing on a button:

```
testWidgets('Making sure that the counter can be updated',
  (tester) async {
  await tester.pumpWidget(const MockWrapper(
    child: CounterAppBody(),
  ));

  final decreaseKey =
    find.byKey(const Key('ElevatedButton-Decrease'));
  final increaseKey =
    find.byKey(const Key('ElevatedButton-Increase'));

  // 0 is the default counter value
  expect(find.text('0'), findsOneWidget);

  // Increasing by 2
  await tester.tap(increaseKey);
  await tester.pumpAndSettle();

  // Finds the counter text and the entry in the history
  // list
  expect(find.text('1'), findsNWidgets(2));

  // Decreasing by 2
  await tester.tap(decreaseKey);
```

```
    await tester.tap(decreaseKey);
    await tester.pumpAndSettle();

    // Only the counter has negative values, the history
    // doesn't
    expect(find.text('-1'), findsOneWidget);
});
```

Using keys is a very common testing strategy to easily get a reference to a specific widget in the tree. When there are multiple widgets of the same type, it may be hard to get a reference to the specific widget you want without keys. In this case, we could have also done this:

```
final decreaseKey = find.byType(ElevatedButton).first;
final increaseKey = find.byType(ElevatedButton).last;
```

No keys have been used here because we know that this widget only has two buttons. But what if we needed to add one more ElevatedButton later? We would need to update the test and keep track of which widget is at a certain index. This would be very time-consuming and unstable. Maintaining tests is not ideal, so we want to make sure we only update them in the case of breaking behavior changes.

> **Note**
>
> If you forget to call tester.pumpAndSettle() after a gesture or anything that requires multiple frames to advance, you may not get the desired result because not all the scheduled frames may be executed.

Now that we have both smoke and behavior tests for our widget, we're just missing a **golden test**. A golden test is a *screenshot* of a widget, stored in a PNG file, that is used later to detect visual changes. Let's take a look at how they work:

1. When you write a golden test, make sure that you run the first test with flutter test --update-goldens. Note that a new PNG image will be generated alongside your test file.

2. Run subsequent tests using flutter test (without the update-goldens flag) so that Flutter will compare the golden image (the *screenshot* you took of the widget) with the widget that's being rendered in the tree. If they don't look the same, the test will fail.

You should use golden tests to produce a master image of how a widget must always look in the UI. Not only will this be helpful for the developer, but it makes UI change testing much easier. In this way, you won't have to manually look for colors, shadows, border radiuses, elevations, and more because Flutter will compare the image with the actual widget.

Let's write a golden test for our CounterAppBody widget:

```
testGoldens('CounterAppBody no history - golden', (tester)
  async {
  final builder = GoldenBuilder.column()
    ..addScenario(
      'No history',
      const SizedBox(
        width: 400,
        height: 400,
        child: MockWrapper(
          child: CounterAppBody(),
        ),
      ),
    );

  await tester.pumpWidgetBuilder(
    builder.build(),
    surfaceSize: const Size(400, 460),
  );
  await screenMatchesGolden(tester,
    'counter_app_body_no_history');
});
```

Here, we're taking a screenshot of the `CounterAppBody` widget when it's first created, meaning that no buttons have been tapped. For this reason, we expect the counter to be at zero and the number history to be empty. The `testGoldens` method and the `GoldenBuilder` class come from the `golden_toolkit` package we depend on. Running `flutter test --update-goldens` will result in the following output:

Figure 9.1 – Golden of the CounterAppBody widget, without history

This is now considered the source of truth by Flutter for what the widget has to look like when rendered. For example, if you try to change the text color and run `flutter test`, you will get an error because the widget won't produce the result that's shown in the preceding screenshot.

> **Note**
> You should only use the `--update-goldens` flag when you need to refresh a golden test or generate a new one. Regular tests, such as the ones you set up in CI, should always use a simple Flutter test command.

To make our widget test even more valuable, let's make a golden test of the body, even when the counter has been tapped. To do so, we are going to use the `customPump` parameter, which allows us to execute some more code before generating the image:

```
testGoldens('CounterAppBody with history - golden',
  (tester) async {
  final builder = GoldenBuilder.column()
```

```
      ..addScenario(
        'With history',
        const SizedBox(
          width: 400,
          height: 400,
          child: MockWrapper(
            child: CounterAppBody(),
          ),
        ),
      );

    await tester.pumpWidgetBuilder(
      builder.build(),
      surfaceSize: const Size(400, 500),
    );

    await screenMatchesGolden(
      tester,
      'counter_app_body_with_history',
      customPump: (tester) async {
        final increaseKey =
          find.byKey(const Key('ElevatedButton-Increase'));

        await tester.tap(increaseKey);
        await tester.tap(increaseKey);

        await tester.pumpAndSettle();
      },
    );
  });
```

Thanks to the `tester.tap(increaseKey)` calls, we can alter the widget's status and since we're making them in `customPump`, we're ensuring they're executed right before the image is generated. Here, we've used a key to get a reference to the increase button.

This is the final result:

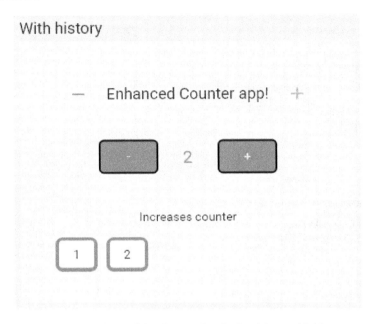

Figure 9.2 – Golden of the CounterAppBody widget, with history

As you can see, golden images (the widget's *screenshots*) don't perfectly match the result you'd get when running on a device. They are very similar, but not identical, and that's fine! They're used by the testing framework to look for changes, not to aesthetically please the developer.

Golden tests can also be made without the `golden_toolkit` package but they don't look as good due to the weird default font. For example, we could test `CounterAppBody` with Flutter's native commands:

```
await expectLater(
  find.byType(CounterAppBody),
  matchesGoldenFile('counter_app_body_no_history.png'),
);
```

Golden files are still generated in the usual way – that is, by appending the --update-goldens flag. While fully functional, the image is not very valuable:

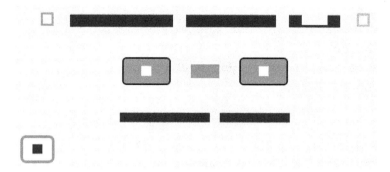

Figure 9.3 – A golden file created with "vanilla" Flutter commands

This result is expected! By default, Flutter uses a font called **Ahem**, which shows squares instead of actual characters and there is no way to change the font. The golden file still works as expected but it would be nicer if the font had human-readable characters.

That's the reason why we decided to use golden_toolkit! It has the useful loadAppFont() method, which loads the Roboto font and applies it to the golden image. This method is called every time a golden test is run because, at the end of the *Writing tests for the Flutter app* section, we created the flutter_test_config.dart file:

```
Future<void> testExecutable(FutureOr<void> Function()
  testMain) async {
  // Do something before ANY test executes here

  // Then run the actual test body
  await testMain();

  // Do something after ANY test executes here
}
```

You must name the file in the same way we did. The Flutter testing environment looks for this configuration file at startup. The testMain callback is the body of the test function you've written, so you can add custom test setups before or after executing it. Make sure that you call await testMain(); somewhere in this method; otherwise, the tests won't run correctly!

Now, let's see how we've tested the `AppTitle` widget, which doesn't have behaviors (no buttons to be pressed, no listeners, no actions to be triggered, and so on). Here, we've just made a smoke test and created a golden test:

```
testWidgets('Making sure that the widget is rendered',
  (tester) async {
  await tester.pumpWidget(const MockWrapper(
    child: AppTitle(),
  ));

  expect(find.byType(AppTitle), findsOneWidget);
  expect(find.text('Enhanced Counter app!'),
    findsOneWidget);
  expect(find.byType(Icon), findsNWidgets(2));
});


testGoldens('AppTitle - golden', (tester) async {
  final builder = GoldenBuilder.column()
    ..addScenario(
      'AppTitle widget',
      const SizedBox(
        width: 300,
        height: 60,
        child: MockWrapper(
          child: AppTitle(),
        ),
      ),
    );

  await tester.pumpWidgetBuilder(
    builder.build(),
    surfaceSize: const Size(300, 120),
  );
  await screenMatchesGolden(tester, 'app_title_widget');
});
```

We're not showing all the tests we've written for the other widgets of the app since the strategy is always the same. When you need to test a widget, try to follow this guideline:

1. Write a smoke test first.

2. Test the behavior of the widget or class by writing widget or unit tests, respectively.

3. Create at least one golden test per widget.

Now that we've provided an overview of the testing phase, we can start using GitHub to set up our repository and automatically execute our workflows using GitHub actions!

Exploring GitHub actions and repository quality

To keep our project healthy, we want to make sure that our tests always pass and the best Dart/Flutter guidelines are respected. We cannot always make these checks manually since it would be error-prone, time-consuming, and not systematic – after all, we're all human and we can forget about tasks!

For this reason, we are going to set up a CI configuration in GitHub that systematically performs a series of checks in our code. GitHub actions, as the name suggests, are a series of actions that automate your workflows. In our case, we will use a Flutter action to install the framework in our server and another action to check the code coverage.

Let's get started!

Creating the GitHub workflow file

We need to create a folder called `.github` at the root of the repository, which is the same place where the `.git` file is located. If you check our online repository, you'll see a variety of content:

*   A `pull_request_template.md` file, which ensures that contributors will automatically see the template's contents in the pull request body. This is very nice to have because it provides consistency across PRs and helps the user submit the request with the data you expect.

*   An `ISSUE_TEMPLATE` folder, which ensures that contributors will automatically see the template's contents in the issue body. This generally contains a few questions the user should answer to provide as many details as possible.

*   A `workflows` directory, which contains the GitHub actions we want to use.

You aren't forced to define a pull request and an issue template, but they are really helpful for external contributors, so consider them! However, the main focus is on the `workflows` folder, where we're going to create a `project_ci.yml` file with the following content:

```yaml
name: chapter_9_ci


on:
  push:
    branches:
      - main
      - develop
  pull_request:
    types:
      - opened


jobs:
  verify_project_health:
    name: Chapter 9 project - CI
    runs-on: windows-latest
    defaults:
      run:
        working-directory: chapter_9
    steps:
      - name: Checkout
        uses: actions/checkout@v2
      - uses: subosito/flutter-action@v1.5.3

      - name: Installing dependencies
        run: flutter pub get

      - name: Making sure the code is formatted
        run: flutter format --set-exit-if-changed .

      - name: Making sure the analyzer doesn't report
          problems
        run: flutter analyze --fatal-infos --fatal-warnings
          && flutter pub run dart_code_metrics:metrics
```

```
      analyze lib test

  - name: Runing unit and widget tests
    run: flutter test --coverage

  - name: Making sure that code coverage is at least 95
    uses: VeryGoodOpenSource/very_good_coverage@v1.1.1
    with:
      path: chapter_9/coverage/lcov.info
      min_coverage: 95
```

And that's it! In particular, the workflow we've just created does the following:

1. First, it uses the `on` directive to determine when the workflow should start. In this case, we want the actions to run whenever we push to the `main` or `develop` branches or whenever a new PR is created.

2. The operating system that the goldens for this project have been created in is Windows. Since the output PNG images of `flutter test --update-goldens` have slightly different results based on the OS that's used (but still notable), we need to run our tests in the same OS. For example, if you generated goldens on a Linux machine and then you tested them on Windows, they would fail. This is expected because different OSs create images using different strategies so, even if it's subtle, there are differences and Flutter detects them.

3. We must install the popular `subosito/flutter-action` action to make the Flutter commands available in the environment.

4. We must make sure that the code is formatted as Flutter would expect using the `--set-exit-if-changed .` flag, which breaks the workflow in case one or more files haven't been formatted.

5. We must ensure no analysis errors are found. Since we are using the `dart_code_metrics` package, we also need to run its specific analysis command to ensure that we follow the guidelines we declared in the `analysis_options.yaml` file.

6. We must run our tests using the `--coverage` flag because we want to make sure that our code is covered above a certain threshold.

7. Finally, we can use the `VeryGoodOpenSource/very_good_coverage` action, which ensures that code coverage is above a certain value and raises an error otherwise.

You just need to place a YAML file in `.github/workflows` to create an automated workflow using GitHub actions. Every time we open a PR or we push to either `main` or `develop`, these checks run and report their statuses in the dedicated GitHub tab. For example, if we forget to format our code or there were some analysis warnings we forgot to clear, the action will fail:

Figure 9.4 – A failing GitHub action report

When the code is formatted, no analysis warnings will be found and all of the tests will pass, so the action will be completed:

Figure 9.5 – A successful GitHub action report

Now, let's learn about code coverage and why it matters!

Code coverage

Everyone agrees that writing a lot of tests is important for various reasons, some of which are as follows:

- They may save hours of debugging and maintenance.
- They help prevent undesired regressions.
- They provide productivity increases because there is no longer a need to manually test features over and over again.
- Tests provide documentation for how the code must behave.

The more tests we write, the less risk we are at of encountering bugs. Code coverage is a testing metric that describes the number of lines of code we have validated using tests. 100% code coverage indicates that our *entire* code has been verified by at least one test, so we can expect a lower chance of encountering bugs.

In Android Studio, for example, you can easily see the code coverage of your Flutter project by clicking on the shield next to the **Run** button. This will show you the coverage percentages of the files in each directory:

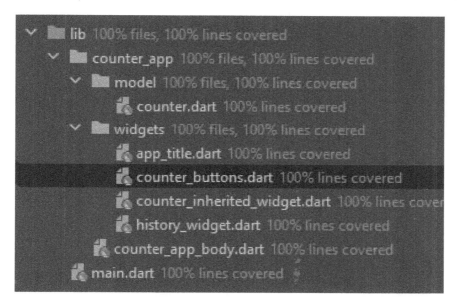

Figure 9.6 – Coverage report in Android Studio

As you can see, our project has 100% code coverage. VS Code has various coverage plugins you can get from its store that allow you to nicely see the coverage percentages as well.

As you've already seen in the workflow file, we can also use the `flutter test --coverage` command to manually generate the coverage report in an LCOV file. The GitHub action we're using parses this file and makes sure that code coverage is higher than 95%.

The LCOV file format is not human readable, so we need some external tools, such as `codecov.io`, to generate a nice visual report. Alternatively, macOS and Linux systems have the `genhtml` utility, which produces a nice HTML overview of the coverage. Run the following command in your Flutter project root directory:

```
genhtml coverage/lcov.info --output-directory=coverage
```

You will find an `index.html` page in the `coverage` folder, which also has clickable paths so that you can see its details in a very user-friendly way:

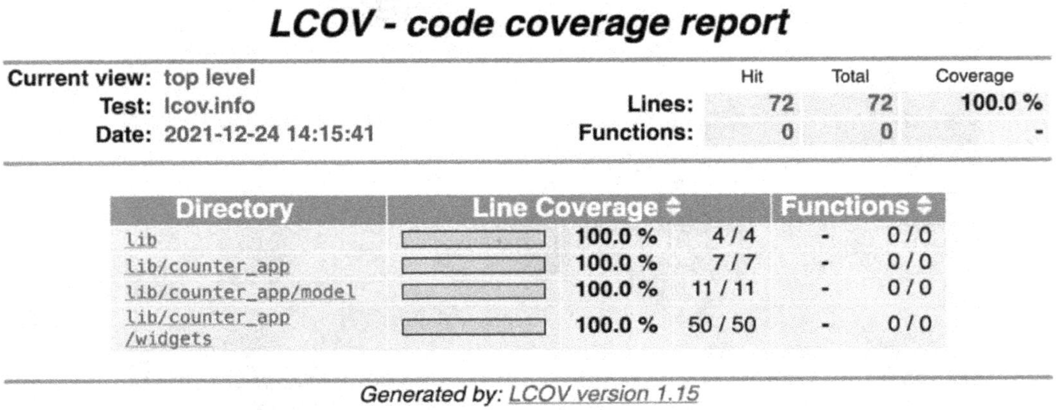

LCOV - code coverage report

Current view:	top level		Hit	Total	Coverage
Test:	lcov.info	**Lines:**	72	72	**100.0 %**
Date:	2021-12-24 14:15:41	**Functions:**	0	0	-

Directory	Line Coverage ⬍			Functions ⬍	
lib	100.0 %	4 / 4	-	0 / 0	
lib/counter_app	100.0 %	7 / 7	-	0 / 0	
lib/counter_app/model	100.0 %	11 / 11	-	0 / 0	
lib/counter_app /widgets	100.0 %	50 / 50	-	0 / 0	

Generated by: LCOV version 1.15

Figure 9.7 – The HTML coverage report that's generated by the genhtml tool

In real-world applications, 100% code coverage may be really hard to achieve, but you should aim for the highest value possible.

Summary

In this chapter, we learned what smoke and golden tests are, as well as how to test behaviors with unit or widget tests. We also learned what golden files are – that is, PNG images that are used as a reference on how a widget must look. These must be tested against the same operating system.

Then, we created an automated workflow that checks our app's formatting, whether we follow various analyzer rules, and ensures that all our tests run correctly. We also made sure that the overall coverage is above a certain threshold using the `VeryGoodOpenSource/very_good_coverage` action.

You can also check your code coverage locally by either using your IDE or the `genhtml` tool, which produces a series of HTML pages.

That's all, folks! In this book, we tried to cover the most important Flutter topics, hoping they will help improve your skillset. We hope you've enjoyed reading this book as much as we did writing it!

Further reading

To learn more about the topics that were covered in this chapter, take a look at the following resources:

- **Testing Flutter apps**: `https://docs.flutter.dev/testing`
- **golden_toolkit package**: `https://pub.dev/packages/golden_toolkit`
- **GitHub actions**: `https://docs.github.com/en/actions`

Index

Packt.com

Subscribe to our online digital library for full access to over 7,000 books and videos, as well as industry leading tools to help you plan your personal development and advance your career. For more information, please visit our website.

Why subscribe?

- Spend less time learning and more time coding with practical eBooks and Videos from over 4,000 industry professionals

- Improve your learning with Skill Plans built especially for you

- Get a free eBook or video every month

- Fully searchable for easy access to vital information

- Copy and paste, print, and bookmark content

Did you know that Packt offers eBook versions of every book published, with PDF and ePub files available? You can upgrade to the eBook version at packt.com and as a print book customer, you are entitled to a discount on the eBook copy. Get in touch with us at customercare@packtpub.com for more details.

At www.packt.com, you can also read a collection of free technical articles, sign up for a range of free newsletters, and receive exclusive discounts and offers on Packt books and eBooks.

Other Books You May Enjoy

If you enjoyed this book, you may be interested in these other books by Packt:

Flutter for Beginners – 2nd edition

Thomas Bailey, Alessandro Biessek

ISBN: 978-1-80056-599-9

- Explore the core concepts of the Flutter framework and how it is used for cross-platform development
- Understand the fundamentals of the Dart programming language
- Work with Flutter widgets and learn the concepts of stateful and stateless widgets
- Add animation to your app using animated widgets and advanced animations techniques
- Master the complete development lifecycle, including testing and debugging
- Investigate the app release process to both mobile stores and the web

Flutter Cookbook

Simone Alessandria, Brian Kayfitz

ISBN: 978-1-83882-338-2

- Use Dart programming to customize your Flutter applications
- Discover how to develop and think like a Dart programmer
- Leverage Firebase Machine Learning capabilities to create intelligent apps
- Create reusable architecture that can be applied to any type of app
- Use web services and persist data locally
- Debug and solve problems before users can see them
- Use asynchronous programming with Future and Stream
- Manage the app state with Streams and the BLoC pattern

Packt is searching for authors like you

If you're interested in becoming an author for Packt, please visit authors.packtpub.com and apply today. We have worked with thousands of developers and tech professionals, just like you, to help them share their insight with the global tech community. You can make a general application, apply for a specific hot topic that we are recruiting an author for, or submit your own idea.

Share your thoughts

Now you've finished *Cross Platform UIs with Flutter*, we'd love to hear your thoughts! Scan the QR code below to go straight to the Amazon review page for this book and share your feedback or leave a review on the site that you purchased it from.

https://www.amazon.in/review/create-review/?asin=1801810494&

Your review is important to us and the tech community and will help us make sure we're delivering excellent quality content.

www.ingramcontent.com/pod-product-compliance
Lightning Source LLC
Chambersburg PA
CBHW060537060326
40690CB00017B/3517